The best indicators of the thermal history of a basin are commonly found in its shales—kerogen and conodont color, reflectance of vitrain, and clay mineralogy.

Total organic content of a shale is very sensitive to the original circulation of its muddy basin, biogenic productivity, and influx of fine terrigeneous or carbonate mud.

Preservation of organics in a muddy basin requires a density stratification to inhibit vertical mixing and minimize oxidation.

Shales are an important primary source of oil and gas.

USES

Shales and their clays have served man well from his earliest time (pottery and adobe) to today, where their uses for modern culture and technology are vast—bricks, cosmetics, drilling muds, insulators, printing inks, medicines, toothpaste and fossil energy to list but a few specific uses. Below are some general headings of how we use clays and shales today.

Absorbents
Bleaching agents
Bond for foundry sand
Ceramic products
   Lightweight aggregates
   Porcelain
   Pottery
   Refractories
   Structural products
Extenders
Gas sources
Landfills
Heavy metals and mineral deposits
Oil source
Pigments
Suspending agents

Paul E. Potter
J. Barry Maynard
Wayne A. Pryor

Sedimentology
of Shale

Study Guide and Reference Source

With 154 Figures and a Colored Insert

Springer-Verlag
New York Berlin Heidelberg Tokyo

PAUL E. POTTER
J. BARRY MAYNARD
WAYNE A. PRYOR
H.N. Fisk Laboratory of Sedimentology, Department of Geology, University of Cincinnati, Cincinnati, Ohio 45221, USA

Library of Congress Cataloging in Publication Data
Potter, Paul Edwin.
  Sedimentology of shale.
  Bibliography: p.
  Includes index.
  1. Shale. I. Maynard, James B., joint author.   II. Pryor,
Wayne Arthur, joint author.   III. Title.
QE471.15.S5P67   552'.5   79-14807

Second printing, 1984.

9 8 7 6 5 4 3 2

ISBN 0-387-90430-1
          Springer-Verlag   New York   Berlin   Heidelberg   Tokyo
ISBN 3-540-90430-1
          Springer-Verlag   Berlin   Heidelberg   New York   Tokyo

Preface

We wrote *Sedimentology of Shale* primarily because we lacked a handy, reasonably comprehensive source of information and ideas about shales for students in our sedimentology program. It was also our feeling that the time for shales to receive more study had finally arrived. *Sedimentology of Shale* also seems very timely because today more sedimentologists are interested in shales. Certainly in the last five years the pace of shale research has noticeably quickened because the role of shales as important sources of oil, gas, heavy metals and as a long understudied part of the earth's geologic history has been recognized. Noteworthy developments include the elucidation of the importance of trace fossils in shales, the discovery of thick sequences of overpressured shales in regions such as the Gulf Coast (which have important implications for hydrocarbon migration and faulting), the extension of the principles of metamorphic facies to the realm of low temperature diagenesis by study of the organic matter in shales, and shales as ultimate sources for mineral deposits.

Accordingly, we decided it was timely to write a book on shales. In one respect, however, ours is an unusual book. Most books in geology are produced after one or two decades of progress have been made in a field and attempt to summarize and evaluate that progress. Our book looks in the foreward direction; instead of summarizing what is well understood about shales, much of our effort has gone into identifying *approaches* to shale problems that we think will be fruitful, and in addition we have emphasized areas of uncertainty.

Chapter 1 provides an overview of the major aspects of shales; including sedimentary processes; physical, chemical and biological properties of shales; and the distribution of shales in modern and ancient basins. Chapter 2, the question set, is in effect a self-study guide, one that not only asks questions but also provides the motivation of *why* the question should be asked. In other words, Chapter 2 is a short manual on *how to study* shales. Chapter 3 is an annotated and illustrated guide to much of the literature. Thus the three chapters are study and reference guides and should provide a basis for students and professionals alike to discover the many interesting, diverse facets of shales. Ultimately, we hope *Sedimentology of Shale* will promote new models for the study of shales.

We have been aided by many, both in North America and abroad. It is a pleasure to acknowledge this important help: P. B. ANDREWS, New Zealand Geological Survey, Canterbury, New Zealand; GEORGE S. AUSTIN, New Mexico Bureau of Mines and Resources, Socorro, New Mexico; J. W. BAXTER, Illinois Geological Survey, Urbana, Illinois; P. BITTERLI, Geol.-Palaont. Institut, Basel, Switzerland; K. BJÖRLYKKE, Universiteti Oslo, Oslo, Norway; ANDREW BODOCSI, University of Cincinnati, Cincinnati, Ohio; CARLOS ALFREDO BORTOTUZZI, Universidade Federal do Rio Grande do Sul, Porte Alegre, Rio Grande do Sul, Brazil; A. BOSELLINI, Universita di Ferrara, Italy; KEES A. DE JONG, University of Cincinnati, Cincinnati, Ohio; M. and C. D. DUMITRIU, Ecole Polytechnique de Montréal, Montréal, Qúebec, Canada; S. DŻULYŃSKI, Instytut Geografii Pau, Krakow, Poland; G. EINSELE, Universität von Tübingen, Fed. Rep. of Germany; K. A. ERIKSSON, Texas Christian Univ., Dallas, Texas; ROBERT W. FLEMING, U.S. Geological Survey, Denver, Colorado; HANS FÜCHTBAUER, Ruhr-Universität Bochum, Bochum, Fed. Rep. of Germany; R. N. GINSBURG, University of Miami, Miami, Florida; R. D. HARVEY and ROBERT CLUFF, Illinois Geological Survey, Urbana, Illinois; DONALD E. HATTIN, Indiana University, Bloomington, Indiana; A. S. HOROWITZ, Indiana University, Bloomington, Indiana; PHILLIP H. HECKEL, University of Iowa, Iowa City, Iowa; WARREN D. HUFF, University of Cincinnati, Cincinnati, Ohio; E. KRINITZSKY, U.S. Corps. of Engineers, Vicksburg, Mississippi; GUNAR LARSEN, Aarhus Universitet, Aarhus, Denmark; SVEN LAUFELD, University of Alaska, Fairbanks, Alaska; FREDRICH LIPPMANN, Mineralogisch-Petrographisches Institut der Universität, Tübingen, Fed. Rep. of Germany; DAVID L. MEYER, University of Cincinnati, Cincinnati, Ohio; GEORGES MILLOT, Université de Strasbourg, Strasbourg, France; HAYDN H. MURRAY, Indiana University, Bloomington, Indiana; HARUNO NAGAHAMA, Geological Survey of Japan, Hisamoto-cho, Kawasaki-shi, Japan; SAMUEL H. PATTERSON, U. S. Geological Survey, Reston, Virginia; FRANCIS J. PETTIJOHN, The Johns Hopkins University, Baltimore, Maryland; BRUCE PURSER, Université de Paris-Sud, Orsay, France; H.-E. REINECK, Senkenberg Institut, Wilhelmshaven, Germany; ROBERT V. RUHE, Indiana University, Bloomington, Indiana; H. H. SCHMITZ, Bundesanstalt für Geowissenschaften und Rohstoffe, Hanover, Fed. Rep. of Germany; ADOLPH SEILACHER, Universität von Tübingen, Fed. Rep. of Germany; JAMES T. TELLER, University of Manitoba, Winnipeg, Manitoba, Canada; WILLIAM A. THOMAS, Georgia State University, Atlanta, Georgia; H.A. TOURTELOT, U. S. Geological Survey, Denver, Colorado; JOHN E. WARME, Rice University, Houston, Texas; and MARCELO R. YRIGOYEN, Esso Argentina, Buenos Aires, Argentina.

C. C. M. GUTJAHR, Koninklijke/Shell Exploratie en Produktie Laboratorium, Rijswijk, Nederlands; K. O. STANLEY, Ohio State University, Columbus, Ohio; and WINIFRIED ZIMMERLE, Deutsche Texaco-Aktiengesellschaft, Wietze, Fed. Rep. of Germany all graciously loaned us thin sections of shales for use in Figure 2-13.

ROBERT CLUFF, RICHARD D. HARVEY and ARTHUR W. WHITE, Illinois State Geological Survey, Urbana, Illinois read parts of the manuscript as did HAYDN MURRAY, Indiana University, Bloomington, Indiana, DAVID L. MEYER, of the University of Cincinnati and ROY C. KEPFERLE, U. S. Geological Survey. We also thank MICHAEL LEWAN and EDWARD PITTMAN, Amoco Production Research, Tulsa, Oklahoma, who also read the manuscript.

LINDA P. FULTON, Exxon Co., U.S.A., New Orleans, Louisiana graciously contributed to the annotated bibliography.

We wish to thank NEIL SAMUELS, PAUL LUNDEGARD, VICTOR VAN BEUREN, RONALD BROADHEAD, RICK TOBIN, and MICHAEL LEWAN, graduate students at the University of Cincinnati, in addition to our typists, JEAN CARROL, MARSHA JONES, DEBBIE MOORMAN, and WANDA OSBORNE. MRS. RUTH SCOTT kindly redrafted some of the figures. JAMES WILLIAMS drafted most of the new figures. We are especially indebted to our departmental librarian, RICHARD SPOHN, for his many hours of considerate help. In addition, we thank the Department of Geology for its support and also the U. S. Department of Energy and the Morgantown Energy Technical Center, whose contract to study the Devonian Shale sequence of the Appalachian Basin helped focus our interest on shales.

Finally, we extend our appreciation to our publishers, Springer-Verlag, and especially to DR. KONRAD F. SPRINGER for his continued encouragement. EILERT ERFLING skillfully guided Sedimentology of Shale to publication.

Cincinnati, Ohio   PAUL EDWIN POTTER
Summer 1980   J. BARRY MAYNARD
                                                             WAYNE A. PRYOR

Contents

Scott W. Starratt
Dept. of Paleontology
U. C. Berkeley
Berkeley, Ca. 94720

Possibly many may think that the deposition and consolidation of fine-grained mud must be a very simple matter, and the results of little interest.   However, when carefully studied . . . it is soon found to be so complex a question, and the results dependent on so many variable conditions, that one might feel inclined to abandon the inquiry, were it not that so much of the history of our rocks appears to be written in this language.

H.C. Sorby

On the Application of Quantitative Methods to the Study of the Structure and History of Rocks, 1908, pp. 190–191.

CHAPTER 1
OVERVIEW

To see the whole of any subject is to see its future.

INTRODUCTION

The study of shale—and we use shale as the generally accepted class name for all fine-grained argillaceous sediment, including mud, clay, and mudstone—can be approached from many points of view (Fig. 1.1), but most of the emphasis has generally been placed upon mineralogy and geochemistry, a possible exception being the study of microfauna in Tertiary and Mesozoic shales. As a consequence, much more is known about sandstones and carbonates and even evaporites than shales. As sedimentologists we have long had an interest in shales and are acutely aware of how far their study lags behind that of most other sediments. All too commonly shale has been the "interbedded" and "taken for granted" matrix between lithologies of greater scientific or economic interest, in spite of the fact that shale forms more than 60% of the world's sediments.

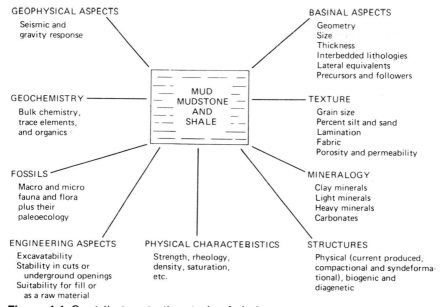

GEOPHYSICAL ASPECTS
Seismic and
gravity response

BASINAL ASPECTS
Geometry
Size
Thickness
Interbedded lithologies
Lateral equivalents
Precursors and followers

GEOCHEMISTRY
Bulk chemistry,
trace elements,
and organics

MUD
MUDSTONE
AND
SHALE

TEXTURE
Grain size
Percent silt and sand
Lamination
Fabric
Porosity and permeability

FOSSILS
Macro and micro
fauna and flora
plus their
paleoecology

MINERALOGY
Clay minerals
Light minerals
Heavy minerals
Carbonates

ENGINEERING ASPECTS
Excavatability
Stability in cuts or
  underground openings
Suitability for fill or
  as a raw material

PHYSICAL CHARACTERISTICS
Strength, rheology,
density, saturation,
etc.

STRUCTURES
Physical (current produced,
compactional and syndeforma-
tional), biogenic and
diagenetic

Figure 1.1 Contributors to the study of shale.

The present attitude of many sedimentologists toward shales is well expressed by their typical representation of shales in vertical profiles which give much detail about the bedding type and thickness, directional structures, grain size, and fossil content of sandstone and carbonates but show mudstones and shales as *solid black*—a structureless, uninteresting, "matrix" interbedded and interlaminated between much more informative lithologies. It has been the experience of many sedimentologists that shales are hard to work with: they are very fine grained and lack the well-known sedimentary structures that are so useful in sandstones. Therefore many feel that work with shale has a low reward/investment ratio. However, is this because of the nature of shale, or is it because we lack readily applicable tools and models to study them? We and, indeed, an increasing number of sedimentologists believe that shales deserve much more attention and, when properly studied, can yield valuable additional insight into the origin of many sedimentary basins.

The natural question to ask is *why* the study of shales has lagged behind that of sandstones, carbonates, and even evaporites, coal, and sedimentary iron formation. Certainly the foremost reason is that shales, as a rule, are of lesser immediate economic interest. Beyond this practical reason, however, are at least four others:

1. Until very recently, it was almost impossible to identify and study *single particles* in most shales, especially the clay minerals, which predominate.
2. Many of these single particles that can now be defined by the scanning election microscope (SEM) have, unlike most quartz or skeletal carbonate grains, a complex history that to some degree reflects source area, possibly the depositional environment, and very probably postdepositional burial.
3. We are only beginning to recognize and interpret the equivalent of the "vertical environmental profile" that has been so spectacularly successful in the study of sandstones and carbonates in the last 10 years.
4. We rarely have an idea of the paleocurrent or paleocirculation systems that have existed during the deposition of most shales.

Of the above four factors that have retarded the study of shales, we believe our inability to *isolate, study, and write the history of single particles,* as has been done so well for most of the framework grains of sandstones and carbonates, is by far the most important. In other words, the complete history of a sedimentary deposit is far better based on the *summation of the histories of its individual particles than on measurement of bulk properties, such as chemical composition and clay* mineralogy. Such measurements average not only the different compositions of many different particles of diverse origins but also combine the effects of source area, depositional environment, and postdepositional change. In short, if sandstones and carbonates could only be studied by a listing of their bulk mineralogy and chemistry, it is clear that we would know little about them. However, assume for a moment that we could, indeed, obtain a "particle history" for shales. What would be the second task? Surely it would be finding an analog to the vertical profile, which for carbonates and sandstones integrates many diverse features into a coherent, useful environmental interpretation, based on bedding, texture, mineralogy, bioturbation, and fossils, and possibly even some geochemical parameters. The equivalent profile for shales will be based primarily on bedding (types, thickness, and degree of perfection), on bioturbation (kinds and abundance), on fossil content (kinds and abundance), and finally on the amount and type of organic matter. It seems to us that these four, along with supplementary evidence from associated lithologies, shale body geometry, bounding contacts, and position in the basin, are essential in developing a better environmental perception of shales. We suggest bedding, bioturbation, fossils, and organic geochemistry as the most useful because they most closely reflect the primary depositional processes of mud deposition. To this we would add information about paleocirculation based on evidence from directional structures of the minor sandstones and carbonates that are commonly associated with most shales or possibly from the orientation of silt, fragments of wood, and/or elongate fossils, such as graptolites, as was done very early by Ruedemann (1897) for the Ordovician Utica Shale of New York.

SOURCES OF MUD

Terrigenous mud consisting of clay minerals as well as fine quartz and feldspar and detrital micas is mostly generated at the earth's surface by the erosion of preexisting muds, mudstones, and shales. We feel this is so, because shale alone forms 60% of the sedimentary section; moreover, sedimentary rocks themselves cover most of the earth's surface (Way 1973, p. 75). The ultimate source is the weathering of silicates formed at high pressures and temperatures and therefore mostly unstable at the earth's crust (Fig. 1.2). Another probable source, at least during the earth's major periods of glaciation, is abrasion by continental ice sheets and production of rock flour. Other sources include volcanic dust (perhaps a major source during the earth's earliest days) and dust from the deflation of continental deserts. Organisms that pulverize and ingest sediment are a very minor source of terrigenous mud.

Clay minerals form from the weathering of primary minerals in the following general way:

$$H^+ + \text{primary mineral} \rightarrow \text{intermediate clay mineral weathering}$$
$$\text{products} + \text{solutions} \rightarrow \text{gibbsite} + \text{solutions}$$

For instance:

$$H^+ + \text{potash feldspar} \rightarrow (\text{sericite}) \rightarrow \text{montmorillonite} \rightarrow$$
$$\text{kaolinite} \rightarrow \text{gibbsite}$$

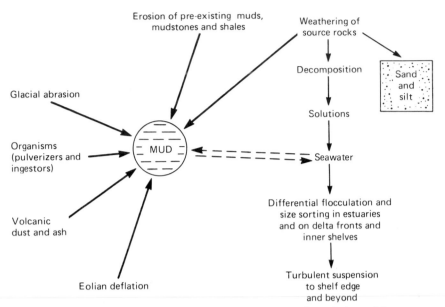

Figure 1.2 Sources of mud and major processes of mud deposition.

H^+ + muscovite $\rightarrow$ illite $\rightarrow$ montmorillonite $\rightarrow$ kaolinite $\rightarrow$ gibbsite

H^+ + biotite $\rightarrow$ (chlorite) $\rightarrow$ vermiculite $\rightarrow$ montmorillonite $\rightarrow$
$\qquad\qquad\qquad\qquad$ Al-smectite $\rightarrow$ kaolinite $\rightarrow$ gibbsite

H^+ + glass $\rightarrow$ gels (allophanes) $\rightarrow$ montmorillonite $\rightarrow$
$\qquad\qquad\qquad\qquad$ halloysite $\rightarrow$ kaolinite $\rightarrow$ gibbsite

Moreover, such minerals as gibbsite, the smectites, and kaolinite can also form in the void spaces of soils as *precipitates* from soil water. The continual flux of water through a weathered zone makes it an ideal site for the production of clay minerals in an open system, in contrast to clay mineral alteration in shales after deposition, which is very largely a closed system.

Climate is very important in clay genesis, as are source rocks. Minimal clay minerals are produced, for example, from quartzite exposed to arid climate weathering (a very chemically stable, low-porosity host with minimum flux of solutions), whereas modern volcanic terrains exposed to tropical weathering produce the maximum volume of clay minerals (unstable glass and minerals in very porous hosts that have a high flux of water and high temperature). The high relief of such volcanic terrains adds to clay production, because it continually exposes new rock to weathering. A good brief summary of clay genesis, from which much of the above has been abstracted, is that of Millot (1978).

Production of mud, as that of all land-derived sediment, is maximal from source areas that have both high relief and high rainfall; together these provide *high potential energy* and the *high kinetic energy* needed to transport solids oceanward. For example, southeast Asia, because it has both high relief and high rainfall, supplies a very large fraction, about 80%, of the yearly suspended sediment carried to the world ocean (Holeman, J.N. 1968 p. 737–747). Mud supplied by big rivers, in contrast, is minimal when their hinterland is low lying and heavily forested. Here soil materials remain for long periods in the zone of weathering so that the final products of weathering are largely solutions and only small volumes of the most stable minerals, such as quartz, kaolinite, or gibbsite, are carried seaward. Another setting that provides minimal mud is a low-lying arid region with few rivers. Still another is a well-watered, low-lying carbonate terrain, which yields chiefly solutions rather than solids. Relief, rainfall, vegetation, and source rocks are all factors that affect mud production. As we shall see, most clay particles, when they finally reach the world ocean, are probably deposited as flocculates and aggregates near its margins.

SUSPENSION, TRANSPORTATION, DEPOSITION, AND EROSION

Fine-grained terrigenous clastic particles are usually transported in hydraulic suspension from their source terrains to their depositional basins and are probably deposited there as aggregates. As with the transport of bed-load sand, however, very much remains to be learned about the essential details of transportation and deposition of muds because both theory and observational data are far from complete. For readers of French, Migniot's (1977) summary of the action of currents, waves, and wind on the transport of sediment is valuable. Below, we briefly examine some of the major controls affecting the suspension, transport, deposition, and erosion of mud.

For suspension transport, the upward components of fluid turbulence must exceed a particle's fall velocity, w. To see this better, consider the equation developed by Vanoni (1941, p. 608) for suspension in rivers:

$$-\phi \frac{\partial c}{\partial y} = cw$$

where ϕ is defined as the transfer coefficient of the suspended particles, $\partial c/\partial y$ is the concentration gradient, c is the concentration (number of particles per unit volume), and y is depth. In a steady state, the rate of downward particle settling, cw, equals the rate of upward particle movement given by $-\phi \partial c/\partial y$. The transfer coefficient of suspended particles, ϕ, is assumed to be proportional to the upward components of the fluid's turbulence (transfer coefficient of fluid momentum) so that the more turbulent the flow, the greater the ϕ and the greater both the concentration and size of particles that can be suspended in it. In rivers, turbulence is proportional to flow intensity (measured by discharge, average velocity, boundary shear stress, stream power, etc.) and is generated chiefly by the river's hydraulically rough boundary of sand grains and bedforms. In standing water, turbulence is generated by wave motion, by thermally induced and suspension currents, and by shear between different water masses as well as by tidal and wind-driven currents. McCave (1971) studied mud deposition in the North Sea and explained it—after fully acknowledging its complexity—using only a few of the above factors, chiefly the suspended concentration c and wave action and near-bottom current velocity, which controls ϕ. He concluded (pp. 94–95) that where mud concentration was low, wave activity and tidal currents greater than 40–50 cm/s inhibited its accumulation, but where suspended mud concentration was high, mud accumulated independent of wave activity.

   The size of a particle that can be carried by a particular flow is expressed by its *fall velocity*, which for spherical particles is given by Stokes Law:

$$w = \frac{\Delta\rho \; d^2 g}{18\mu}$$

where $\Delta\rho$ is the density difference between particle and fluid, g the acceleration of gravity, d the grain diameter, and μ the dynamic viscosity of the fluid. Water strongly affects this equation because viscosity is very temperature dependent—cold water markedly increases viscosity and thus reduces w, all other factors being equal. According to Migniot (1968, p. 595), for spheres with a density between 2.5 and 2.6 settling in pure water at 20°C, d and w are related, as a first approximation, by:

$$d = w^{0.5}$$

where d is measured in microns and w in microns per second. According to this equation, a sphere 8 μm in diameter will fall at 64 μm/s. For quiet, nonturbulent, isothermal water, the values of w given for selected sizes of clay and silt (Table 1.1) show the very slow rates at which they sink. Most clay mineral particles are platy, rather than being spherical, and their fall velocities are even slower than those shown in Table 1.1. Lerman (1979 p. 263–267) provides a comprehensive discussion of settling velocities for particles with different shapes and his Table 6.3 gives correction factors to be used for different shapes. Disc- and cap-shaped particles, the shapes of most individual clay minerals, settle only about half as fast in quiet, nonturbulent water as spheres of equivalent volume. In reality, however, lakes, seas, and oceans do not lack turbulence and currents, so settling rates must be even slower. Because deposition of mud does, in fact, occur, Pryor (1975, p. 1253) concluded that *most mud is deposited as aggregates and floccules rather than as individual particles.* McCave (1975, p. 500), after a comprehensive review of the literature and an analysis of the vertical mass flux in the ocean, also concluded that the aggregation of suspended particles in the ocean is the norm and occurs very rapidly.

   Deposition of mud from such suspensions may occur when: (1) water evaporates or sinks into the ground in playa lakes and in the ephemeral

Table 1.1 Fall Velocities of Spherical Particles in Non-Turbulent Isothermal Water Based on Stokes Law

Diameter	Time to fall 1 m			Fall velocity
(μm)	Days	Hours	Minutes	(cm/s)
60	0	0	5	0.223
30	0	0	30	0.0558
16	0	2	0	0.0139
8	0	7	48	0.00349
4	1	6	0	0.00087
2	5	6	0	0.000217
1	21	10	0	0.000054
0.5	89	0	0	0.000013

flood basins of rivers; (2) mud settles one particle at a time in still water in lakes, seas, and oceans; (3) suspended muds are aggregated and pelletized by organisms; and (4) most commonly, the particles flocculate and settle as aggregates that are much larger than single grains, and so require much higher velocities for suspension. These aggregates, floccules, or pellets behave much like sand or silt grains and can be transported and deposited in various bedforms.

How do the two aggregation processes, biologic pelletization and inorganic flocculation, occur? Clay mineral particles, unlike quartz grains, can be flocculated and deposited by a change in water chemistry. The most important example of this effect in nature is the increasing rate of deposition of mud with increasing salinity. A river carrying mud in suspension tends to deposit much of it as floccules on encountering high-salinity sea water near its mouth, so that many suspended clays are quickly deposited near the shoreline. Migniot (1968, pp. 595–597) has summarized the variables that affect the rate of deposition of floccules. Size of the individual mineral particles has little effect; the increasing tendency of smaller grains to aggregate offsets their smaller fall velocity. Migniot further shows that deposition rate is strongly affected by concentration of the particles and, as mentioned, by the salinity. Functional relations for these two variables are discussed by Van Olphen (1978, Chapt. 7).

Floccule formation and size are enhanced by high suspension concentrations and agitated, turbulent waters. Under these conditions there is a high rate of interparticulate collisions. The floccules increase in size by composite aggregation and by the attraction of individual clay particles to the surfaces of floccules. Suspended silt and fine sand particles are often trapped within floccules. Depending upon mineralogy, salinity, and collision rate, floccules may range in size from a few tens of microns to over 700 μm. They are roughly spherical and have specific gravities ranging from 1.4 to more than 2.0 (Krone, 1962). Initially, floccules have a high water content, which is reduced by synomesis and compaction as the interstitial water is expelled.

An additional geologic variable that has been studied in some detail in recent years is the mineralogy of the particles. Apparently kaolinite flocculates much more readily than illite or mixtures of illite and smectite so that there tends to develop, subject to modification by wave action, a seaward change from kaolinite to illite plus smectite (Parham 1966; Edzwald and O'Melia 1975).

One should be aware, however, that not all clay mineralogists agree that differential flocculation explains the lateral mineralogical distribution of clays. For example, Gibbs (1977) studied the Amazon mud wedge that extends from the Amazon's mouth to the Orinoco delta, a distance of almost 1400 km, and concluded that smectite increased relative to other clays downcurrent from the Amazon's mouth simply because of its finer size. He suggested that possibly organic films or metallic coatings might have inhibited flocculation. Thus flocculation may be predominantly an estuarine process.

There are many marine environments where suspended materials occur in low concentrations and are apparently in equilibrium with the chemistry of the water so that flocculation processes are inhibited. There is increasing evidence that such suspended mud is aggregated by filter-feeding organisms and deposited as fecal pellets or pseudofecal pellets. Organisms, in both shallow marine environments and deep-water basins, are capable of pelletizing large volumes of suspended particles. The pellets sink to the bottom and are transported and deposited as silt- and sand-size grains. Pryor (1975) estimated that just two of the filter-feeding species in Mississippi Sound along the Gulf Coast are capable of pelletizing 12 metric tons of suspended sediments in 1 km²/year and depositing a layer 4.5 mm thick. Little is known about the production rates of argillaceous fecal pellets in other areas, but it may be one of the most important processes for deposition of suspended sediments in interdeltaic and basinal environments.

Another major way mud, silt, and sand are introduced to a basin is by gravity flow. Gravity flow (Middleton and Hampton 1973, Fig. 1) is a transport process that includes turbidity currents, debris flow, grain flow, and fluidized sediment flow (Fig. 1.3). Each of these can introduce mud and sand into a basin and, to some degree, produces distinctive bedding types, which are described in the section "Sedimentary Structures."

Turbidity currents are mixtures of water and detritus, mostly mud, silt, and sand kept in suspension by turbulence, that move downslope under the force of gravity and closely follow the bottom. The mode of sedimentation is settling from turbulent suspension as the turbulence of the current decays. Turbidity currents, most commonly initiated by a slide or slump on an unstable slope, may be of high concentration and carry much sand and even some fine gravel as well as mud, or they may be of low concentration and carry only mud and the finest silt. Much of the mud deposited in the thick shaly basins of ancient geosynclines and continental margins may have accumulated in this manner. The three additional processes of gravity flow—

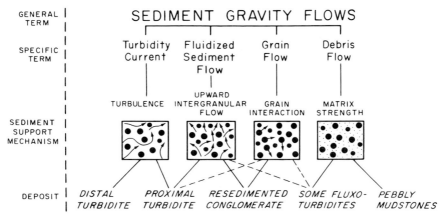

Figure 1.3 Classification of subaqueous gravity flows (redrawn from Middleton and Hampton, 1973, Fig. 1).

debris flow, grain flow, and fluidized sediment—mostly transport mixtures of sand and large clasts and are not, therefore, major factors to be considered in deposition of pure shales. Some sandstone interbedded with pure shales may, however, have been transported by these processes.

From the foregoing it can be seen that mud deposition is virtually independent of water depth. In marine basins and lakes, mud can accumulate at the shoreline wherever it is protected from intense coastal wave power, by a long, shallow shelf, a barrier island or reef, or by coastal vegetation. Mud may also accumulate along many shorelines where the amount of mud in the water column is so high that it inhibits wave action near the bottom, and also retards *inshore* wave power. Conversely, it can occur in protected topographic lows on shelves far from the shoreline and, of course, in deep oceanic basins, large or small. Mud also occurs in lakes and in alluvial valley fills formed by meandering river systems—overbank deposition in temporary flood basins beyond natural levees—and in the interdistributary lakes and bays of low-energy deltas subject to occasional flooding. Irrespective of water depth, mud particles will accumulate in protected basins—*ones that may range from a few centimeters to hundreds of meters deep.*

What can be said of the erosion of a muddy bottom? This depends primarily on the cohesion of the mud, which is largely a function of its water content. Unlike sands and coarse silts, muds have cohesion because of the attraction of broken bonds at the boundaries of their clay minerals and the strength of their water envelopes. As a consequence, a greater flow intensity is required to erode a muddy bottom than a sandy bottom (Fig. 1.4), just as a lesser flow intensity is needed to erode a mud saturated with water than one from which much water has been expelled (Southard 1974, Fig. 2). Bioturbation also is a factor (Young and Southard 1978).

By using the yield strength of the mud on the bottom to define its plastic

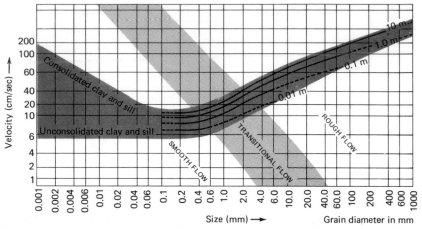

Figure 1.4 Generalized curve for erosion and deposition of mud, silt, and sand modified from Sundborg (1956, fig. 13). Although there are many other similar curves from later studies, all show a broad minimum (inflection) at about 0.2 mm. The wide wedge-like band at the left reflects Sundborg's estimate of the difference in critical velocities needed to ercde clay because of differing plasticities (water contents). Dotted lines and numbers indicate critical velocities at different depths above base.

state, which mostly depends upon its water content, Migniot (1968, p. 613) obtained two equations relating critical or threshold velocity, v_c, expressed in centimeters per second to the yield strength of the mud, τ:

$$v_c = 0.5\tau^{0.5} \qquad \text{for stiff muds with } \tau > 20 \text{ dynes/cm}^2$$

and

$$v_c = \tau^{0.25} \qquad \text{For soft muds with } \tau < 20 \text{ dynes/cm}^2$$

These two equations are significant for sedimentologists, because they show the danger of inferring current competence from only the size of mud clasts when, in fact, the stiffness of the underlying mud is a major factor (i.e., its water content or porosity). Some of the muds studied by Migniot had water contents of 90%−95%.

DEFINITION AND CLASSIFICATION

At this point in our overview, let us turn from our discussion of processes and consider problems of definition and classification, problems that have always been troublesome for muds and shales. Many of these troublesome problems could probably be resolved, if we better understand the major processes that form mud and shale.

Rock classification should enable an investigator to easily categorize the rocks under study and to communicate that information in a concise, understandable way to others. Classification is especially useful for grouping materials into similar identities to derive intrinsic relationships between samples, formations, and facies. The history of sandstone and carbonate classifications clearly shows that descriptive classifications, based on easily observable features (both in hand specimen and in thin section) are generally superior and more widely acceptable than genetic-based classifications. A soundly based description has a significantly longer life span than ever-changing genetic interpretations. As scientists, we know that genetic interpretations should be kept in separate mental compartments from description, but that description should ultimately lead to interpretation. Because classification often shapes the concepts and the ways we study rocks, it is important that a genetic scheme not control description.

By fine-grained terrigenous sediment we mean all rocks and sediments that contain 50% or more of terrigenous, and generally argillaceous, clastic components less than 0.062 mm. In the literature and in common usage, we find such names as lutite, psammite, argillaceous sediment, mud and mudstone, silt and siltstones, shale, argillite, phyllite, and slate. How are these terms defined and currently used? Are the terms precisely defined or are they loosely interchangeable? For a discussion of the history of the term shale, see Tourtelot (1960), who concluded that on a historical basis, *shale* is the generally accepted *class* name for fine-grained rocks and is equivalent to the terms *sandstone* and *limestone,* a usage that we follow here.

There have been several attempts to classify shales, but at present none seems to have found wide acceptance. Could this be the result of the general lack of intense investigation of fine-grained rocks, as was true of carbonates prior to 1950? Careful description-based classifications of both sandstones and carbonates were very few in number prior to modern petrologic studies of these rocks in the 1950s and 1960s.

The classification of fine-grained rocks proposed by Twenhofel (1937) and used by Pettijohn (1975, Fig. 8-1) is based on two variables—state of induration and the relative amounts of silt (quartz?) and clay (mineral?) components. The classification of Picard (1971), emphasizes texture and composition, principally of the silt-sized components, modified by the composition of clay minerals. Picard extended conventional sandstone terminology into the realm of fine-grained rocks; his classification requires thin-section petrographic study. Füchtbauer and Müller (1970, p. 131) also based their classification on state of consolidation and relative amounts of silt (quartz?) and clay minerals. A very recent classification scheme is that of Lewan (1978), who used thin sections to estimate silt content and semiquantitative X-ray diffraction to determine mineralogy. His classification differs from others in that fissility does not play a major role in assigning a name. Lundegard and Samuels (1980) propose a field classification, based on grain size and stratification, that is nearly identical to that in Table 1.2.

Shales, mudstones, argillites, and most other common fine-grained ter-

Table 1.2 Classification of Shale (More than 50% Grains Less than 0.062 mm).
Part A. Nomenclature

Percentage clay-size constituents			0-32	33-65	66-100
Field Adjective			Gritty	Loamy	Fat or Slick
NONINDURATED	Beds	Greater than 10 mm	BEDDED SILT	BEDDED MUD	BEDDED CLAYMUD
NONINDURATED	Laminae	Less than 10 mm	LAMINATED SILT	LAMINATED MUD	LAMINATED CLAYMUD
INDURATED	Beds	Greater than 10 mm	BEDDED SILTSTONE	MUDSTONE	CLAYSTONE
INDURATED	Laminae	Less than 10 mm	LAMINATED SILTSTONE	MUDSHALE	CLAYSHALE
METAMORPHOSED	Degree of metamorphism	LOW	QUARTZ ARGILLITE	ARGILLITE	
METAMORPHOSED	Degree of metamorphism		QUARTZ SLATE	SLATE	
METAMORPHOSED	Degree of metamorphism	HIGH	PHYLLITE AND/OR MICA SCHIST		

rigenous rocks and sediments contain significant amounts of clay minerals and clay-size carbonate, kerogen and silica. It is these clay-size components that often dominate their rock properties, such as density, plasticity, parting, compactibility, swelling, and weathering. Therefore the classification we propose, which elaborates that of Blatt, Middleton and Murray (1971, Table 11.1), recognizes the importance of the clay-size compo-

Table 1.2. Part B. Commonly Used Modifiers

Mineralogy	From petrographic or X-ray diffraction analysis: quartz, feldspar, kaolinite, illite, smectite, etc.; mineralogic modifiers include: Calcareous, carbonaceous, dolomitic, ferruginous, feldspathic, glauconitic, gypsiferous, micaceous, nodular, phosphatic, pyritiferous, quartzose, siliceous, etc.
Carbonate	Calcite ordolomite determined by dilute HCl and stain in the field or by thin section or X-ray; calcareous, dolomitic, sideritic, etc.
Color	Colors from color chart designation;
Induration	Hard, soft, plastic, etc.
Fracture	Conchoidal, hackly, blocky, brittle, splintery earthy, etc.
Grain size	From hand lens or binocular microscope using grain size comparator, percentages of sand–silt–clay fractions
Lamination and bedding	Measured thicknesses; parallel, wavy, lenticular, flaser, disturbed, etc.
Fossil content	Kind, quantity, condition, etc.
Bioturbation and trace fossils	Kind and quantity
Organic constituents	Woody, spores, kerogen, odor, etc.

nents of fine-grained rocks and sediments. We believe shales can be comprehensively, descriptively, and adequately classified on the basis of the following properties: state of induration, relative amounts of clay-size constituents and silt-sized mineral particles, and bedding and laminae thickness, all of which can be readily determined in the field and in thin-section (Table 1.2).

We use the term *shale* as the major class name for fine-grained terrigenous rocks based on the historical arguments presented by Tourtelot (1960) and common usage. A major feature of Table 1.2 is the subdivision based on the relative percentages of the argillaceous component (clay minerals) and the clay-size carbonate, kerogen and silica. This is not only a mineralogic subdivision but also a textural one. As defined by Wentworth (1922, Table 1), the *size* boundary between clay and silt is 4 μm, but most clay minerals are less than 2 μm. In addition, the measurement of sizes in this range is at best difficult, both in the field and in the laboratory. However, the relative amounts of clay versus silt and sand can be determined much more easily. Amounts of clay-size minerals can be roughly estimated in the field, with practice, by hand lense inspection, and the grittiness, plasticity and smoothness to feel and, of course, thin sections can be used to great advantage, especially at the beginning of the study.

The boundary between beds and laminae is also an important distinc-

Table 1.3. Stratification and Parting of Shales

Thickness	Stratification		Parting	Composition	
30 cm	Thin	Bedding	Slabby		
3 cm	Very thin				
10 mm	Thick		Flaggy	Clay and organic content	sand, silt and carbonate content
5 mm	Medium	Lamination	Platy		
1 mm	Thin		Fissile		
0.5 mm	Very thin		Papery		

tion, which subdivides the indurated shale subclasses. In our classification (Table 1.2) the suffix -stone is used to denote layering greater than 10mm (beds) and the suffix -shale to denote layering less than 10mm (laminations). Hence, we restrict the commonly used term siltstone to bedded silt-rich rocks and coin the term siltshale for those silt-rich rocks that are laminated. We propose (Table 1.3) a refinement of the existing *bedding–lamination* schemes (Ingram 1954, McKee and Weir 1953) by providing more subdivisions of lamination and correlating them with parting. Lamination and bedding may be defined by differences of grain size, composition, and/or fabric (arrangement and perfection of orientation by particles). Parting is the tendency of a rock to split along lamination or bedding, a tendency always greatly enhanced by weathering.

Metamorphic processes, especially high temperatures, readily alter shales. Argillites are weakly metamorphosed rocks, firmly indurated without fissility or slaty cleavage and with some of the clay minerals and micas reconstituted to sericite, chlorite, epidote or green biotite. Progressive metamorphism of shales generally results in gradual transition from argillite to slate to phyllite to mica schist with an increase in grain size. Slates have cleavage that is independent of original sedimentary structures. Phyllites are intermediate between slates and schists, and have coarser mica grains than slate. The chlorite, biotite and sericite crystals impart to the phyllites a silky sheen on the cleavage or schistocity surfaces. Mica schists are strongly foliated, with well aligned, large mica crystals and represent high grades of metamorphism.

We believe that this classification will prove to be simple and easy to use in the field and laboratory and at the same time adequate to provide a good basis for the description of shales. Three typical field descriptions follow:

Mudshale: Dark greenish gray (5GY4/1); hard, brittle, platy parting; 5% quartz sand, 35% quartz silt, 60% clay; medium laminae (1.5 mm); nonfossiliferous; and micaceous.

Clayshale: Variegated, 70% light green (5G7/4), 30% moderate red (5R5/4); hard, hackly, flaggy parting; medium laminae (4 mm); 20% quartz silt, 80% clay, small coaly fragments, and small pelecypod impressions; gradational base.

Mudshale: Brownish black (5YR2/1) on fresh surfaces to medium light gray (N6) on weathered surfaces; fissile and brittle; about 70% clay minerals; fine laminae; abundant *Lingula* at top. Some small flattened amoebiform phosphate nodules near top. Sharp basal contact.

The percentages in the above descriptions can be determined in the laboratory by thin section and pipette analysis, and used to refine field estimates.

It is important to remember that classifications are meant to be used and to be helpful. They should not be the "tail that wags the dog." If the data do not fit the classification, it is the classification that must be altered, not the data. Therefore, as investigations of shales increase, evolution of the proposed classification scheme is expectable and desirable.

SEDIMENTARY STRUCTURES

Sedimentary structures have proved to be very useful in the study of sandstones and carbonates, providing us with much insight into their origin, and there is a vast literature (Table 1.4). Using sedimentary structures, information on depositional environments, dispersal patterns and paleocurrents, folding and failure properties, paleoecology, and even some of the geochemistry of diagenesis can be developed. However, very little has been done; usually the sedimentary structures of the interbedded sandstones or carbonates have been described (e.g., Cole and Picard 1975). Do pure shales lack sedimentary structures or are the structures of the interbedded sediments simply much more readily observed and more informative?

Certainly it is true that most primary sedimentary structures occur in silt and sand and are of current-traction origin, whereas most muds seem to be

Table 1.4. Source Materials for the Study of Sedimentary Structures

Angelucci and others (1967)
  Sixty-five figures, most of which are sedimentary structures of sandstones
  of turbidite origin, which are common in many thick shaly basins.
Conybeare and Crook (1968)
  Provides many illustrations of sedimentary structures, mostly in sand and
  sandstones, along with their description and interpretation plus a short
  section on how to analyze sedimentary environments. Although emphasis
  is on sandstones, the same principles also apply to many shaly basins.
Dimitrijević and others (1967)
  A small volume with brief descriptions of 52 beautiful line drawings of
  the inorganic and organic structures of turbidites. Serbian with key ideas
  also in English, French, and German.
Dżulyński (1963)
  Beautifully illustrated treatment of sole marks based on the pioneer work of
  the Polish school, of which Dżulyński is a leader. Useful to all those study-
  ing shales that have some interbedded sandstones and siltstones.
Dżulyński and Sanders (1962)
  Thirty-seven pages of text and 22 excellent plates dealing mainly with sole
  marks that occur on the undersides of sandstones.
Dżulyński and Walton (1965)
  Devoted mainly to sedimentary structures, most of which are directional. A
  well-illustrated summary.
Ginsburg (1975)
  Forty-five papers on tidal deposits, which are largely recognized by their
  vertical sequence of sedimentary structures. Many excellent illustrations.
Gubler and others (1966)
  Contains a short review on stratification and stratigraphic terminology
  followed by a long section (186 Pages) on sedimentary structures. Very
  systematic and complete. Definition, description, measurement, frequency
  of occurrence, origin, and utility are given for each structure. Well illus-
  trated.
Khabakov (1962)
  Probably the first comprehensive picture book of sedimentary structures
  and textures. Most emphasis is on sandstone and carbonates.
Lanteume and others (1967)
  Sixty-one very good plates, mostly of sole marks, with full captions in
  French, English, German, Italian, and Spanish. Full cross index and all the
  previous relevant literature. Essential companion for the field study of turbidite deposits.
Pettijohn and Potter (1964)
  Primarily a picture book prefaced by a short essay on classification and
  followed by a four-language glossary of 360 entries. Scattered structures
  in shales.
Potter and Pettijohn (1977)
  Standard, classic reference on paleocurrents, with many illustrations of
  sedimentary structures. Updated, with over 100 pages of additional text
  and many new illustrations.

Table 1.4. *Continued*

Reineck and Singh (1973)

What they are, how they formed and what they mean—the very essence of what we need to know about sedimentary structures. Mostly about sands and sandstones, but with more on mud than most of the other references in this list.

Ricci Lucchi (1970)

A beautifully illustrated book of primary sedimentary structures, mainly sole marks of flysch sandstones; 170 plates with marginal text. Italian with English–Italian lexicon. Outstanding.

deposited by settling from suspension. Even where mud floccules or fecal pellets are deposited by traction, postdepositional compaction tends to obliterate such structures. Hence it would seem that by far the most common sedimentary structure of "pure" shales—shales free of even minor interbeds of sandstone and carbonate—is horizontal stratification.

Types and Significance of Sedimentary Structures in Shales

It can be argued that horizontal stratification and parting are the *only* primary hydraulic structures in pure shales. By *stratification* we mean layering that is the result of vertical differences in composition, texture, and/or grain fabric. Whatever their origin, however, these vertical differences are manifested by color and hardness in layers of variable thickness. *Parting,* however, is a splitting characteristic of shales, often enhanced by weathering, wherein planes of separation occur between layers.

In Table 1.2 the commonly used classes of stratification and parting are organized, adding to earlier schemes (Alling 1945, Ingram 1954, McKee and Weir 1953) to place more emphasis on shales. Stratification is subdivided into beds and laminae at the traditional 10-cm thickness boundary and these in turn are further subdivided. Parting is divided into the same thickness categories using traditional stonemason's terms: slabby, flaggy, and platy. And what about the term *fissile?* The term *fissile* has not always been used consistently; sometimes it has been used as a loose synonym for parting. Here, however, we follow Alling (1945, p. 753) and define *fissile* simply as a class of parting with thickness between 0.5 and 1.0 mm. Thus the presence or absence of fissility does not define a rock as being a shale.

The thickness of stratification and parting in shales is related to many factors, including rates of sedimentation and compactional state. However, field observations suggest that the most obvious factors are composition, grain size, and fabric. In a very general way, the relationships shown in Table 1.2 hold true: stratification and parting generally decrease in thickness as the relative amounts of clay minerals and organic compounds increase, as

SEDIMENTATIONS-ERSCHEINUNGEN — BODENLEBEN

UNITS	Lithologie	Sediment-Strukturen	Fossil-Orientierung / Einkippung Beobachtungen	Milieu / sessiles Benthos / Lebensspuren

Column headings (Milieu / sessiles Benthos): Stillwasser · Turbul.ber. · Seegangs-β · Ruhen-α · Bromungs-β / Bewong.-β · Brachiop. · rob. Korallen · Stromtac. · Kalkalgen

Lebensspuren facies (top): Skolithos-Fazies · Cruziana-Fazies · Zoophycus-Fazies · Nereiten-Fazies · Spurentiere

1 — Sandstein, Tonschiefer — *keine ebene Feinschichtung* — — Nereitomichnites RI; Diplocraterion
2 — Knollenkalk, Untrein Mergel, teils sandig — — — MA MA; Chondrites (oben) MA
3 — Schwarzschiefer etwas kalkig — ebene Feinschichtung — — Chondrites MA
4 — Knollenkalke und -Mergel, fein geschichtet — regelmäßige Feinschichtung — — MA; Chondrites MA
5 — Dünnbänk. Knollenkalke, gut gebankt — Rippelfeld Schillbänke MA — Lesedecken MA / ⌀ MA — MA MA; Chondrites MA
6 — Wechsellagerung: Kalk u. Tonschiefer, sedim. Bänke — MA / MA — Strömlanata~100%, Lesedecken MA / ⌀ MA — MA MA; Chondrites MA / Zoophycus MA
7 — Tonschiefer, dünne Kalk-sandstein-Bänke — Priele mit Lepidaena-Schill MA — MA Donergelo MA / ② — MA MA, SA SA; Ch. Sterosoma HO, Diplochn. MA
8 — Tonschiefer, dunkel an der Basis Knollenkalk — Convolution Dishe-vers — MA / ⌀ MA — MA SA; Chondr. ③, Rusophycus
9 — Block-Konglomerat — Grob detritische Rinnenfüllung m. Azolli Rippeln u. Block-Konglom. Azolli-Rippeln — ⌀ — HO HO
10 — Kalksandstein, grobkörnig, diabontig — ebene Lamination z.T. winzige Schräg-schichtung / Flow rolls HO — Holorhynchus SA /))22 — HO MA; MA; diplger. : Diplichnites OR / Petecy... podolamus MA, Scolamna MA
11 — Kalksandstein, Tonschiefer — ebene Lamination (u. Rippelschrift) — Fensierten regellos MA — ⌀ MA; SA; HO; Ch. Trichophycus HO

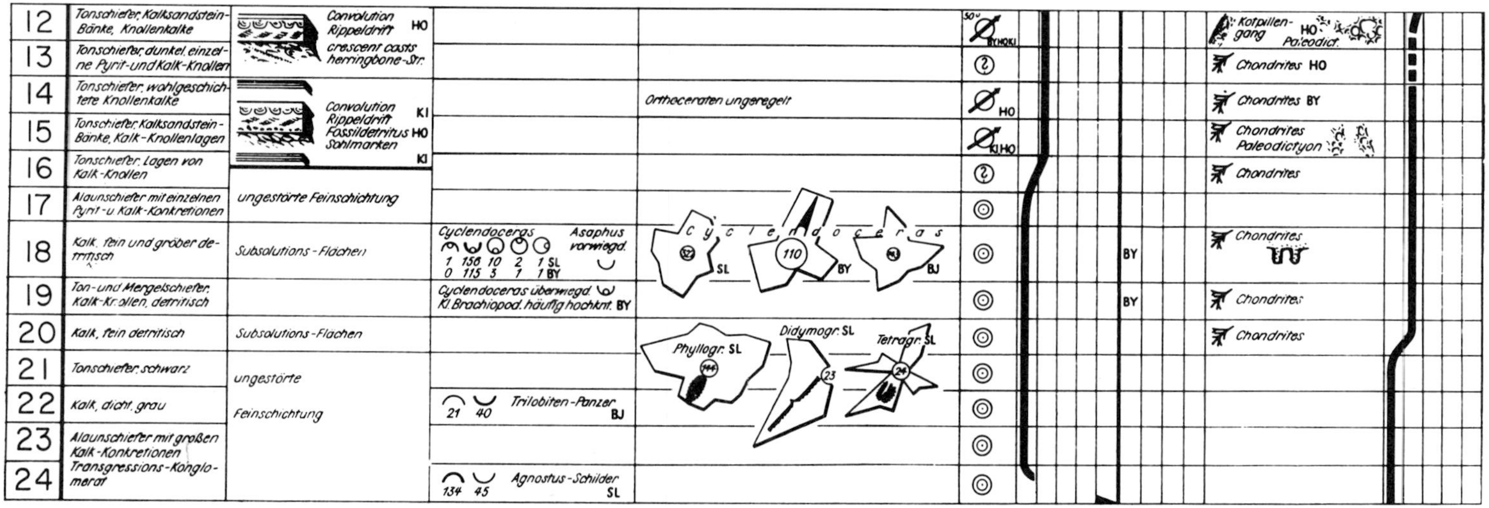

Figure 1.5 Careful integration of lithology, sedimentary structures, body fossils (and their orientation) and trace fossils are the essential basis of most environmental interpretations as was done for this shaly Lower Paleozoic basin in Norway (Seilacher and Meischner, 1964, Table 1). Studies such as this only require careful field observations. English equivalents of German lithologic units: 1) sandstone, shale; 2) nodular limestone, impure, somewhat sandy marl; 3) black shale, somewhat calcareous; 4) nodular limestone and marl, finely bedded; 5) gritty nodular limestone, well bedded; 6) interbedded limestone and sideritic shale; 7) shale with thin beds of calcareous sandstone; 8) dark shale at the base of nodular limestone; 9) coarse conglomerate; 10) coarse, thick bedded calcareous sandstone; 11) calcareous sandstone and shale; 12) shale, calcareous sandstone beds and nodular limestone; 13) dark shale with some pyritic and limestone nodules; 14) well bedded, nodular limestone and shale; 15) shale, calcareous sandstone plus limestone and nodular beds; 16) shale with calcareous nodules; 17) alum shale with some pyrite and limestone concretions; 18) limestone—fine with some grit; 19) gritty shale and marl, limestone nodules; 20) limestone, fine and gritty; 21) black shale; 22) dense, gray limestone; 23) alum shale with large calcareous concretions; and 24) transgressive conglomerate.

the degree of orientation of platy minerals increases, and as the percentages of sand- and silt-sized mineral fragments decrease.

When shales intimately interbedded with sandstones and carbonates, and these are probably the most common, are considered, a wide range of structures is found, and in practice it always pays to study carefully the associated and interbedded lithologies of a shaly basin. A little known but excellent study by two German sedimentologists shows the value of studying sedi-

Table 1.5. Sedimentary Structures in Shale and Their Origin
Part A. Primary Structures

Stratification
  Parallel horizontal
    Episodic suspension in still water
  Massive
    Continuous, rapid sedimentation from suspension or bioturbation
  Parallel discontinuous
    Episodic suspension with some bottom currents
  Lenticular–wavy
    Episodic traction transport with possibly some deposition from suspension
  Varves
    Suspension grading with rapid sedimentation in spring and slow sedimentation in winter
Ripple marks and flaser bedding
  Traction transport of silt, sand, and mud aggregates as ripples with some deposition from suspension
Cross bedding
  Traction transport of sand and silt as large ripples and dunes
Parting lineation
  Traction transport of silt and sand in the "flat-bed" mode
Sole marks
  Bottom scour followed by deposition
Graded beds and Bouma cycles
  Deposition by turbidity currents
Massive sand beds
  Deposition by grain flow
Convolute lamination, dish structures, and fluid escape pipes
  Formed by fluidized sediment flow
Pebbly mudstone and conglomeratic beds
  Deposition by debris flow
Clay clasts
  Local erosion and deposition of cohesive clay layers
Raindrop Imprints
  Subaerial impact by rain drops

mentary structures of all types in shaly basins. Seilacher and Meischner (1964) provided an outstanding example (Fig. 1.5) of how field observation of sedimentary structures and fossils made possible a comprehensive environmental and paleocurrent analysis—one that did not require any laboratory study. Imagine the number of shaly basins that invite similar studies! From a slightly different viewpoint, Reineck (1967) related the sedimentary structures of interlaminated muds and sands from slope, shelf, coastal, tidal, and lacustrine environments to sedimentary processes of muddy basins. These two examples, one ancient and the other modern, show the importance of sedimentary structures in the study of shaly basins.

Below we have organized the sedimentary structures associated with both shales and interbedded shales, sandstones and carbonates into three genetic groups (Table 1.5): (1) those that are primary and most of which are formed by hydraulic processes; (2) those formed after deposition by fluid loss, compaction, and deformational processes; and (3) diagenetic structures formed by chemical processes, some of which form very soon after deposition (synsedimentary) and others, later.

Table 1.5. Part B. Compactional and Deformational Structures

Mud cracks and slickensides
  Dessication and shrinkage either by subaqueous syneresis or subaerial drying
Load casts
  Ball and pillows
    Soft sediment displacement of sands and silts into underlying mud
  Flame structure
    Soft sediment displacement of sands and silts into underlying muds and sliding downslope
Mud lumps and diapirs
  Large-scale upward displacement of plastic mud and shale

Table 1.5. Part C. Diagenetic Structures

Concretions
  Nodules
  Septaria   } Local cementation, commonly early, without major displacement of mud matrix; commonly form around organic nucleus
  Geodes
  Spherulites
Cone-in-cone
  Crystal growth
Crystal casts
  Crystal growth, commonly salt
Color banding
  Probably diffusion and generally obscure, but may be related to weathering

---

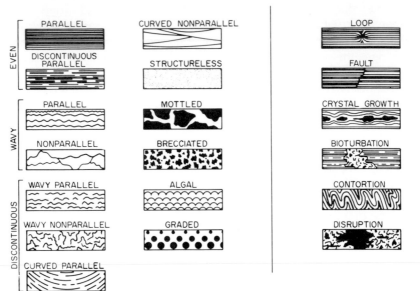

Figure 1.6 Diagrammatic illustration of major primary stratification types and secondary structures in the Eocene Green River Shale of Wyoming (Cole and Picard, 1975, Fig. 2) applies to many shaly basins, especially those with little interbedded siltstone, sandstone or carbonate.

Primary Structures

Primary sedimentary structures are formed by hydraulic processes during deposition of shales and related rocks. They may be enhanced by post-depositional processes, but they are chiefly a record of primary sedimentation. For shales, these processes are flocculation and pelletization, suspension and settling, traction transport, and the various types of gravity flow. The various types of horizontal stratification are of the utmost importance in the vast majority of shales (Fig. 1.6). Campbell (1967 Table 1) also provides a classification of bedding and lamination.

Parallel horizontal stratification is characterized by distinct beds or laminae, with either sharp or gradational boundaries, that are continuous and parallel. Such lamination or bedding represents the most uniform and regular type of depositon that occurs in shales. Parallel, even lamination represents sedimentation in quiet water, where bottom currents are too weak to sculpture the bottom and where clay and only the finest silt are transported. *Discontinuous,* but even, parallel stratification is closely related and represents deposition only slightly less uniform. Here the laminae either terminate abruptly or taper. *Wavy* or *lenticular* stratification may be parallel, nonparallel, or discontinuous and is chiefly formed by small ripples of silt and sand which, when interbedded with mud, are called *flaser bedding.* Compaction may enhance bedding irregularities of hydraulic origin. Traction transport of floccules or fecal pellets may also produce wavy bedding. Wavy bedding of all kinds implies a small but distinct microrelief on the bottom and higher current velocities than parallel stratification. *Discontinuous wavy, nonparallel* bedding is a less orderly arrangement and probably represents a higher level of current action. *Curved parallel* or *curved nonparallel* bedding is locally present in many shales and, although a universal explanation is not at hand, probably represents lateral accretion in channel scours. *Massive* or *structureless* thin bedding or laminae are common and may represent very uniform sedimentation over a short time or may result from the destruction of bedding by burrowing organisms.

These various forms of stratification represent episodic sedimentation, where boundaries are sharp, each bed or lamina separated in time from adjacent ones. Where boundaries are gradational, they represent continuous sedimentation with differential settling rates of various constituents, or variations in material supplied to the basin. Deposition may be from suspension, where individual clay mineral or silt particles or aggregates of clay mineral particles, formed by flocculation and aggregation, settle to the bottom according to Stoke's Law. Parallel horizontal stratification probably represents sedimentation from suspension in relatively quiet water, where bottom currents are too weak to sculpture the cohesive bottom sediment and where silt and clay aggregates are transported in the lower flow regime mode and possibly by episodic precipitation of minerals or organic matter. When bottom current velocities are high enough to erode the bottom sediment or move granular materials in the traction mode, irregular, wavy, or lenticular beds or laminae form microrelief on the bottom.

Beds and laminae are sedimentation units that represent episodic events in the history of deposition. These events may be related to fluctuations in terrigenous sediment supply and may be seasonally controlled (Bradley 1931) or they can also be caused by resuspension and settling during storms or floods. Still another possible factor is water chemistry, which affects organic productivity and controls the precipitation of minor amounts of minerals, such as calcite and gypsum. Algal bedding seen in the carbonates of some calcareous shales is an example of bedding produced by carbonate-trapping organisms that are sensitive to both water chemistry and light.

Varves associated with glacial deposits are classic examples of seasonally controlled sedimentation. By definition, varves are yearly sets composed of a light colored silt layer, deposited rapidly in the summer, grading into a darker organic-rich layer deposited slowly during the winter.

Ripple marks and their internal micro-cross-laminae are sedimentary structures of silt- and sand-sized materials which are common in shales. Rippled silts and sands intercalated with shales are often best exposed on the upper surfaces and display a wide variety of geometries (J.R.L. Allen 1968, Fig. 4.61). Ripples of silt and sand occur interbedded with shales as isolated

Figure 1.7 Starved ripple composed of skeletal carbonate debris in shale of Cincinnatian Series (Ordovician) near Moscow, Clermont Co., Ohio, USA. (Pettijohn and Potter, 1964, Pl. 90B).

or "starved" ripples (Fig. 1.7), as single or multiple layers, and as lenticular layers intimately mixed with shales (commonly called "flaser bedding"). Reineck and Wunderlich (1968, Figs. 2, 3, 4, and 5) show a continuum of rippled sand–shale mixtures that range from starved ripples to lenticular and wavy bedding, with flaser bedding as the end member (Fig. 1.8). Flaser bedding develops when mud fills in the ripple troughs and is then covered by successive ripples. The muds interbedded with rippled silts and sands may result from intermittent slack water conditions, during which clays settle out of suspension, or they may represent the traction transport and deposition of fecal pellets or floccules during intermittent lower current velocities. Some argillaceous fecal pellets have hydraulic equivalences of 0.5-mm diameter quartz sand grains (Pryor 1975, p. 1247–1248).

Ripple crest and micro-cross-lamination orientation (Fig. 1.9) are especially useful in deducing paleocurrent and dispersal patterns in shaly basins, where they are very common. In addition, the rippled sand or silt to shale ratio may be useful in determining water depth in shoreface sequences (Reineck and Singh 1973, Table 20 and Fig. 485).

Cross-stratification, a very common bedding feature of sands and sandstones, is only a minor feature of most shaly basins, except possibly along their margins. Cross-stratified sandstones record relatively high, turbulent flow velocity conditions; therefore their occurrences with low flow velocity shales are records of widely fluctuating current velocities. Episodes of high-

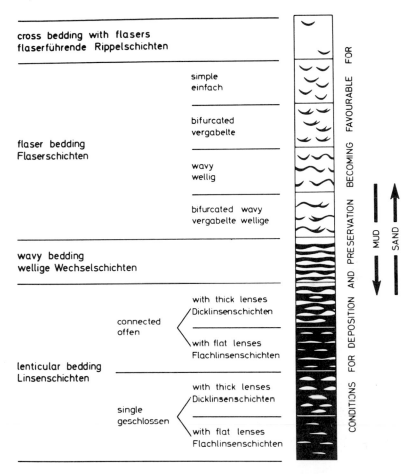

cross bedding with flasers
flaserführende Rippelschichten

simple
einfach

bifurcated
vergabelte

flaser bedding
Flaserschichten

wavy
wellig

bifurcated wavy
vergabelte wellige

wavy bedding
wellige Wechselschichten

with thick lenses
Dicklinsenschichten

connected
offen

with flat lenses
Flachlinsenschichten

lenticular bedding
Linsenschichten

with thick lenses
Dicklinsenschichten

single
geschlossen

with flat lenses
Flachlinsenschichten

CONDITIONS FOR DEPOSITION AND PRESERVATION BECOMING FAVOURABLE FOR

MUD

SAND

Figure 1.8 Terminology for interbedded sequences of mud and ripple sand (Reineck and Singh, 1973, Fig. 164. Republished with the authors' permission.)

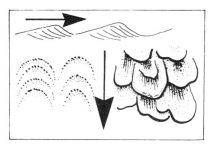

Figure 1.9 Sketch of ripples that indicate direction of paleoflow (Dimitrijević and others, 1967, Fig. 40.) Ripples such as this are common in shaly basins.

Figure 1.10 Parting lineation such as this can occur on some thin sandstones, especially in their upper portions, interbedded with shales. Arrowheads show possible current directions (Dimitrijević and others, 1968, Fig. 33).

velocity flow conditions, wherein cross-stratified sand layers are formed on a mud substrate, can be generated in the marine environment by storm surges, where shallow-water sands are resuspended and carried into the deeper muddy bottom environment, and to a lesser degree by turbidity currents that carry sands into deep-water mud environments. Such currents produce laminated, cross-laminated, ripple bedded, and convolute structures in the sands they deposit. In shallow water, periodic high-velocity conditions can cause quartz or carbonate sand waves to migrate over subaqueous firm mud surfaces (Fig. 1.7). Flocculated muds and argillaceous fecal pellets also can be formed into megaripples and migrating shoals (Krone 1962) and cross-stratified floccule clay sediments have been reported by Pryor and Van Wie (1971) from the Eocene of Tennessee and Kentucky. *Cross-lamination,* inclined foresets less than 10mm. thick, may be far more common in shales than is generally recognized and can occur in sets a few centimeters thick.

Parting lineation, also known as current lineation, is a sedimentary structure of thin-bedded, horizontally stratified siltstones and sandstones that commonly occurs in shaly basins. This structure is seen on parting surfaces by faint, elongate and linear, parallel, stepped ridges of detached laminae (Fig. 1.10). Parting lineation is related to the parallel orientation of elongate grains (McBride and Yeakel 1963), which are aligned in response to paleocurrents. Parting lineation is possibly indicative of both upper and lower flow regime conditions and is found in both shallow-water deposits and deep-water turbidites.

Sole marks are castings of various kinds on the bottom surfaces of quartz or carbonate sandstone and siltstone beds. The forms (Fig. 1.11), include flutes, tool marks (such as bounce, brush, prod, and groove casts), and current crescents (Dzulynski and Sanders 1962) and are formed by filling the relief on mud surfaces with quartz or carbonate detritus. The relief on the mud surface is the result of sculpting a relatively firm mud bottom by hydraulic scour or by cutting grooves, dents, and ridges on it with various tools. All that is needed is a relatively firm mud bottom, a current flowing

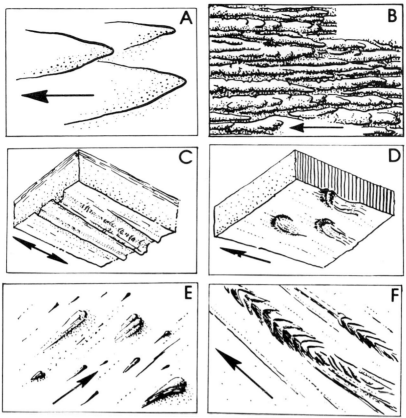

Figure 1.11 Sketches of sole marks: A) flute (blunt end points up current); B) longitudinal furrows and ridges; C) groove marks; D) brush casts; E) prod casts; and F) chevron marks (Dimitrijević and others, 1969, Figs. 3, 12, 21, 23, 28 and 29).

over it with sufficient competence to erode it or pull a tool across it, and a following layer of sand- or silt-sized grains to bury and preserve the resulting relief as a sole cast. These conditions are met in many depositional environments ranging from fluvial, lagoonal, tidal, and shallow marine to deep marine. Sole marks, however, are most commonly found on the bottoms of sandy turbidite layers in relatively deep-water shaly deposits. In shales the texture of sole marks on interbedded siltstones is commonly very fine (Fig. 1.12).

There are, however, many turbidite layers that have very few sole-marked bases. Why are they absent? The major, critical factor in producing sole marks is the cohesiveness of the mud substrate. If it is too soft, mud is simply resuspended by turbidity current flow; however, if the substrate is too firm to be sculptured, no relief will develop. Sole marks require intermittent sedimentation where the mud substrate is exposed for a sufficient period of

Figure 1.12 Low relief flutes and weak grooves on base of thin siltstone in black shale. Blunt end of flutes point upcurrent. Ohio Shale (Devonian), Rocky River Reservation, Cleveland, Cuyahoga Co., Ohio, USA.

time to develop the proper cohesiveness. Hence the absence of sole marks in turbidites either may indicate intermittent sedimentation with very short or very long periods of time between turbidite events or it may indicate continuous sedimentation of hemipelagic muds, where the mud substrate never develops a cohesiveness, with episodic turbidity currents.

"Bouma cycles" (Fig. 1.13) are another feature of turbidity currents. The "ideal" Bouma cycle consists of texture and bedding subdivisions that are the result of changing hydraulic regimes. The top (E) unit is composed of silt and clay, usually weakly bedded, which is called the "pelagic shale" or pelitic interval. Part of this interval may represent deposition from the lutite part of a turbidity current and part may represent normal pelagic mud deposition. Natland (1933), in an early classic paper, and later Hesse (1975, Table 2), discussed the identification and separation of these two possibilities. Few Bouma cycles are "ideal"; most have one or more of the five subdivisions missing. Missing units may be attributed to either the lack of proper sized material in the source or the successive, downcurrent sedimentation of the coarser materials first and the finer materials last. From this pattern, proximal and distal Bouma sequences can be defined and used to

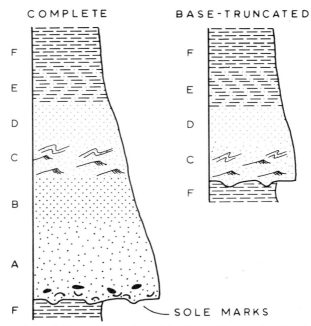

Figure 1.13 Complete and partial Bouma cycles with six and four units. Such cycles are characteristic of the interbedded siltstones and sandstones of turbidite basins and may provide the principal clue to one origin of the great mass of shale in such basins.

reconstruct dispersal patterns (R.G. Walker 1967). The pelitic "E" unit or pelagic shale is nearly ubiquitous and may form the bulk of the turbidite bed in the most distal setting parts of a turbidite basin. Sole marks on the base of the turbidite, parting lineation in the B and D units, and micro-cross-lamination in the C unit are all useful indicators of paleocurrents and dispersal patterns and may be the only indicators in the same shaly units.

Pebbly mudstones and *conglomerate beds* (diamictites, fluxoturbidites, paraconglomerates, mud flows, tilloids, Gerröllton, etc.) can be intercalated with shales and have long been troublesome to understand, particularly in thick marine shale sections. The principal problem is how to explain such a juxtaposition.

Pebbly mudstone units are composed of various mixtures of boulder, gravel, sand, silt, and clay; beds may range to more than 50 m thick. The fine-grained matrix (mud) may range from less than 10% to more than 90% and the clasts may be of any composition available, including locally derived soft-sediment clasts. Most pebbly mudstones intercalated with shales are associated with turbidite facies and are formed by debris flow (Hampton 1972), which is the flow of a viscous mass of cohesive mixtures of granular solids, clay, and fluids down a slope in response to gravity. Pebbly

Figure 1.14 Solitary conglomerate bed in Cretaceous shale near Salvador along fault-bounded margin of Recôncavo Basin, Bahia, Brazil. (Photograph courtesy of Petrobras.) Solitary conglomerate beds such as these in shale are evidence of deposition along an unstable margin of a basin—the cobbles and pebbles slid downslope into muddier and deeper water.

mudstones of debris flow origin and isolated conglomerates in shaly basins are normally associated with the most proximal facies of a turbidite sequence (Walker 1978). The conglomeratic beds that are interbedded with some shales (Fig. 1.14), which have long puzzled many investigators, were probably emplaced by mass movements of this kind.

Clay clasts (clay galls, clay pebbles, ripup clasts, etc.) are sand- to boulder-sized clasts of cohesive mud. They can occur within shale units or, more commonly, within silt and sand units interbedded with shales. Very typically such clasts are oriented. The clasts may be angular to rounded, discoid to spherical, and also deformed; may occur as framework constituents or single clasts dispersed in a matrix; and may be randomly oriented, be imbricated, or have oriented long axes. Clay clasts are locally derived either by reworking dessicated, mud crack-generated fragments or by the hydraulic erosion of cohesive mud laminae or beds. Some very large clasts are produced by gravity slumping or as "pull-aparts" during mass movement of cohesive muds on slopes and have been called "wild flysch" and olistostromes.

Raindrop imprints are one of those interesting, but seldom seen sedimentary structures. They are small, 0.5–1.5 cm in diameter, symmetrical or asymmetrical, shallow, rimmed depressions, produced by the impact of raindrops or hailstones on a soft, subaerial mud surface. However, they may be confused with craters made by air or gas or water escaping from sediment, either subaerially or subaqueously.

Compactional and Deformational Structures

Compaction and deformation structures are formed during or immediately after deposition of a sediment. These "soft-sediment" structures are records of events and conditions in the environment between depositional events. They are formed by gravitational movements, either downslope or vertical compression; by density differences; by intergranular fluid movement; or by dessication processes. Muds and muddy sands are particularly prone to "soft-sediment" deformation, because pore water pressures do not dissipate rapidly in low-permeability muds.

Mud cracks in shales have always fascinated geologists and have traditionally been used as indicators of subaerial exposure; indeed they are often called "sun cracks" (Fig. 1.15). However, mud cracks can form not only subaerially, but also by several subaqueous processes. Mud cracks are usually formed by shrinkage, or dessication of muds. Dessication is a process wherein interstitial waters are removed and mineral grains become more densely aggregated. This can result in the development of polygonal or rectilinear networks of cracks. Interstitial water can be lost from muds by evaporation in a subaerial environment, but just as commonly it can move out of a muddy substrate, under water, by the process known as syneresis (Jüngst 1934). Syneresis is a continuation of the flocculation process wherein clay mineral particles move toward each other by mutual attaction, either electrochemical or gravitational, and the interstitial water is forced out and general shrinkage takes place. Investigations of this colloidal phenomenon by White (1961) and Burst (1965) have shown that syneresis cracks form where clays have been deposited by flocculation and that the kinds of clay minerals and salinity of the interstitial waters are important in development of the cracks. White has also shown that syneresis is responsible for much of the slickensiding found in claystones, such as the underclays of coal beds. Jüngst and White also noted the development of small circular pits on clay surfaces, formed by escaping streams of water, and suggested that such pits may be mistaken for raindrop imprints.

No satisfactory criteria have yet been developed to distinguish subaqueous from subaerial mudcracks. Mudcracks can also form simply by downslope creep of a muddy substrate. Cracks of this origin are seldom polygonal, usually being rectilinear and oriented perpendicular to slope direction. Cracks in muddy sediments are often filled in by overlying sand layers and are expressed as sole marks on the interbedded sandstones. The

Figure 1.15 Mudcracks are common in marine shallow water and continental shaly basins, although few are as spectacular as these from a playa in Esmeralda Co., Nevada, USA.

cracks are often, but not always, polygonal; they taper downward to a pinchout and may extend through several underlying layers.

Load casts, such as ball and pillows, flow rolls, and flame structure (Fig. 1.16), are formed by downward gravitational displacement of soft sands and silts into underlying soft muds. These structures protrude downward from the base of sandstone beds. In some shaly basins the ball and pillows, flow rolls, and flame structures are asymmetrical in a uniform direction, indicating a downslope displacement and can be used as paleoslope indicators. Kuenen (1958, p. 18) has experimentally demonstrated that uneven loading can lead to foundering of unconsolidated sands into a quasiliquid substrate, producing these types of structures.

Convolute lamination and *synsedimentary folds* are usually found in thin-bedded shale, siltstone, fine sandstone sequences, and turbidites. The laminae and beds are deformed into small anticlinal and synclinal folds,

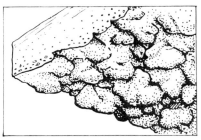 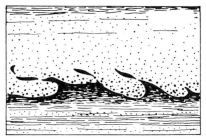

Figure 1.16 Load casts (left) and flame structure (right) are common in thin interbeds of sandstone and/or carbonate in shaly basins (Dimitrijević and others, 1968, Figs. 30 and 31).

often recumbent and sometimes faulted, whereas the overlying and underlying beds remain undisturbed. Some soft-sediment deformation of glacial clays and varves is probably the result of ice shove processes. Where the convolutions and folds are asymmetrical, they may represent failure down a slope and so are useful for determining paleoslope directions. Expulsion of water from sands can also result in dikes of sand being forced into adjacent muds.

Mud lumps and *mud diapirs* are important large-scale loading features and are often associated with low-density, undercompacted muddy sediments, especially in the thick sections of rapidly subsiding basins. Because they are early structures, they can be expecially important hydrocarbon traps. Mud or shale diapirs may range from a few to hundreds and even thousands of meters wide.

Diagenetic Structures

Diagentic sedimentary structures are generally looked upon as geologic oddities and desk-top curios. As yet, most such structures in shales have not been systematically studied and related to the major facies of shaly basins. This is in part because the structures have either been described as oddities or studied as isolated and abstract "geochemical factories." A few exceptions to this generality include Hallam's (1967) studies of siderite nodules, Weeks' (1953, 1957) investigations of carbonate concretions in shales, the classic paleoecologic study of the Pennsylvanian–Mazon Creek fossiliferous siderite concretions by Zangerl and Richardson (1963), and Raiswell's (1971) work on the timing of concretion growth.

Diagenetic sedimentary structures are those that form after deposition, either early or late. Many, if not most, appear to form very early, probably because the mud still has appreciable permeability and strong chemical gradients exist between the mud and the overlying water. Hence, diagenetic structures are records of the geochemical character of ancient substrates. In addition, because some of these structures form very early, they can be useful in estimating degree of compaction in shales. Shales seem to be the most common host rock of many diagenetic structures, probably because muddy sediments have high fluid contents, interstitial fluids migrate slowly through them, their clay mineral constituents are chemically reactive, and oxygen levels are generally low.

The most common diagenetic sedimentary structures in shaly sequences are the concretionary ones listed in Table 1.3: *concretions, nodules, septarian concretions, spherulites,* and *geodes.* Todd (1903) and Pettijohn (1975, pp. 462–482) have discussed the classification, shape, sizes, composition, and possible origins of these diagenetic structures. Shapes range from spherical and oblate spherical to irregular, and sizes range from microspherules several microns in diameter to enormous concretions up to 10 m in diameter (Fig. 1.17). They may be composed of chert, calcite, aragonite, dolomite, siderite, hematite, limonite, marcasite, pyrite, gypsum, anhydrite, and barite as the more common minerals. Many concretions and nodules are associated with organic compounds and, indeed, nuclei of animal or plant fossils or trace fossils are common (Reiskind 1975). Late-formed concretions seem to occupy zones of high permeability. Outcrop weathering can dissolve parts of concretions and produce geodes.

Precipitation of the mineral matter can either displace the host rock, occupy only the pore space in the host rock, or occur in voids and open fractures, or mineral matter can precipitate syngenetically at the sediment–water interface (Voight 1968). The character of bedding and lamination of the host shales will often be the best clue to determining the mode of precipitation (Raiswell 1971).

The precise mechanisms for the precipitation of the various mineral segregations in shale host rocks are incompletely known, but an association with organic compounds seems to be a common denominator. Weeks (1957) thought that the fossil-bearing calcareous nodules in shales of the Magdalena Valley in Colombia were formed upon the decomposition of organisms' corpses. Ammonia was produced as the tissue decayed, creating a microenvironment with a high-pH. Such an increase in pH would precipitate calcium carbonate from the pore fluids and form a nodule around the organic matter. The uncrushed character of the fossils clearly shows that the nodules were formed prior to compaction (Weeks 1953). Experimental verification of this process has been presented by Berner (1968), and a similar model has been proposed by Raiswell (1976). Zangerl and Richardson (1963, Fig. 27) interpreted the very fossiliferous siderite nodules of two Pennsylvanian black shales in northern Illinois to have formed in the plastic bottom muds of low-oxygen, poorly circulated pond environments.

Pyrite and marcasite nodules and spherulites are common in shales. They may be associated with or replace fossil plant and animal fragments, but normally replacement is so pervasive that evidence for such associations is hard to establish.

Geodes in shales may form around fossil nuclei, such as crinoid calyxes

Figure 1.17 Large hollow concretion of calcium carbonate in Huron Member of Ohio Shale (Devonian) at Copperas Mountain, 3.9 miles east of Bainbridge, Ross Co., Ohio, USA. Concretions in most shales tend to be strongly associated with selected facies. They may also occur in very thin stratigraphic units, which may be very widespread and thus be useful marker beds. More mapping of concretionary facies by sedimentologists is needed.

and brachiopods, but for the great majority it is difficult to establish a possible nucleus. Because the host rock seems to be displaced, many of these structures are thought to have formed after at least some burial (Hays, 1964 and Fisher, 1977).

Many concretions and nodules occur in well defined zones and can be well oriented with respect to paleocurrent directions (Colton, 1967 and Jacob, 1973).

Cone-in-cone layers are another distinctive diagenetic structure in shales and consist of sets of nested, interpenetrating cones. Such cones usually occur as discontinuous layers, a few millimeters to a few centimeters thick, but may also enclose concretions. They are almost always calcite but may have originally been aragonite and seem to be associated with the more organic-rich shales. Any original structure within the beds is obliterated by recrystallization (Usdowski 1963, Gilman and Metzger 1967, Franks 1969, W.S. MacKenzie 1972).

A promising avenue of research on diagenetic minerals is stable isotope

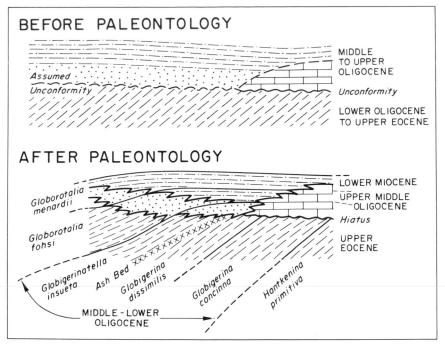

Figure 1.18 Contrasting interpretations of a basin before and after paleontologic study. (Redrawn from Stainforth and others, 1975, Fig. 3.)

matter (the precursor to petroleum) to the basin and terrestrial plant detritus
(spores, pollen, and cuticle) contribute lignin-rich organic matter (precursor to natural gas) to the basin. Hence the type of organic matter (marine or
terrestrial) and the type of hydrocarbon deposit are a function of the organis-
mal association. An excellent recent text on how organisms produce petroleum is the one by Tissot and Welte (1978) and more on the organic
geochemistry of shales is found in the "Mineralogy and Geochemistry" section of this chapter.

The quantity of organic matter in a basin is a function of biologic productivity, which is controlled by optimum conditions of light, temperature, and
mineral nutrients, especially phosphorus and nitrate, as well as the preservation of organic matter after the death of the organism. In shaly basins, especially those where circulation is restricted and sedimentation rates are
moderate to high (lagoons, estuaries, deep basins, and continental slopes),
oxygenation rates are low and conditions are very favorable for the accumulation and survival of lipid-rich and lignin-rich organic matter.

Marine phytoplankton, zooplankton, and bacteria were the major contributors to the organic matter of basins from Cambrian through Jurassic times,
whereas terrestrial plants began contributing organic matter during the
Devonian and became major contributors from the Cretaceous to the
Recent.

Organisms also contribute significantly to our perception of the environment of deposition and, indeed, when properly studied, their contribution can far surpass that of physical and chemical sedimentology. This is true because different communities of organisms commonly have different requirements for light, salinity, turbidity, temperature, amount of free oxygen, type of substrate, and wave and current energy. Of these, turbidity, character of the substrate, and amount of free oxygen are the most important for the vast majority of organisms living in muddy environments.

Both body fossils and trace fossils contribute to our perception of the environments of deposition of shaly basins and, for that reason, knowledge of both is important to sedimentologists who study basins. To work effectively with body fossils for age determinations—and certainly for their ecologic significance—requires the special skills of considerable paleontologic training. Therefore, it is nearly always best—perhaps necessary is a better word—for most sedimentologists to work closely with either macro- or micropaleontologists. Micropaleontology is especially important for shales because microfossils may be the only ones present. In the Question Set (Chapt. 2) the reader will find a fuller discussion of references to both micropaleontology and paleoecology as well as questions that should be posed. However, the ability to recognize the use of trace fossils should be a part of the equipment of all sedimentologists and we therefore emphasize it here.

Biogenic structures of shales and interbedded sediments are formed by the activities of organisms, both plants and animals, during and shortly after deposition. These are manifested in sediments by differences in composition, texture, and fabric. Biogneic structures are records of organic activities in response to ecologic factors, such as salinity, water depth, oxygen level, temperature, light, wave and current levels, water turbidity, substrate character, sedimentation and erosion rates, food supply, and population density. These are all environmental factors that we need to know about in order to interpret the origin of the sediments in which the structures are found. The paleontologic aspects are just as important: biogenic structures may be the only records of soft-bodied organisms and they also record the behavioral patterns of extinct organisms. These observations add fundamental knowledge to our understanding of the fossil record. Biogenic structures are also kinds of fossils that are indigenous to the rocks in which they are found, a biocenosis. The sedimentologic, stratigraphic, paleoecologic, and paleontologic aspects of biogenic structures are discussed in depth by Frey (1975), Reineck and Singh (1973), and Crimes and Harper (1970, 1977).

Biogenic structures fall into three categories. They may be distinct, morphologic structures, known as *trace fossils* or lebensspuren. Dense, interpenetrating, generally indistinct trace fossils are referred to as *burrow-mottled* textures. Nonspecific mixing of the sediment is generally known as *bioturbation*. Trace fossils may be classified either following the genetic basis of Seilacher (1953) or in one of several descriptive schemes, such as

epichnial   epichnial
epichnial groove   ridge   load impression

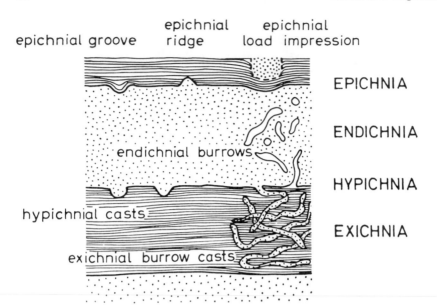

EPICHNIA

ENDICHNIA

HYPICHNIA

EXICHNIA

Figure 1.19 Terminology of trace fossils according to Martinsson (1965, Fig. 12).

that presented by Martinsson (1965). The toponomic terms used in the latter scheme (Fig. 1.19) are descriptive, simple, and therefore most useful for investigators not trained in ichnology (the study of trace fossils). Taxonomy and genetic interpretations can be added to the descriptive base. Like the sole marks on turbidite beds, trace fossils are often best displayed as surface markings on sandstone or siltstone beds interbedded with shales.

Because most trace fossils have distinct morphologies, which are susceptible to taxonomic classification (Fig. 1.20), the trace fossils themselves have been given generic and specific names, even though the organisms that have made them may be unknown. Different organisms that have similar habits and activities may produce similar, if not identical, trace fossils. This has led to the behavioral or ethologic classifications, such as that proposed by Seilacher (1953) and modified by Simpson (1975). There are six generally accepted categories of behavioral trace fossils—crawling traces (Repichnia), grazing traces (Pascichnia), feeding traces (Fodinichnia), dwelling traces (Domichnia), resting traces (Cubichnia), and escape traces (Fugichnia). Crawling, grazing, and resting traces are usually bedding plane features, either hypichnia or epichnia, whereas feeding, dwelling, and escape structures are endichnia or exichnia.

Trace fossil and bioturbation assemblages are very common in shaly sequences and are extremely useful in deducing sedimentologic and paleoecologic characteristics. The relative abundance of trace fossils or the intensity of bioturbation can reflect rates of sedimentation. Abundant, concentrated trace fossils and bioturbation may represent slow sedimentation

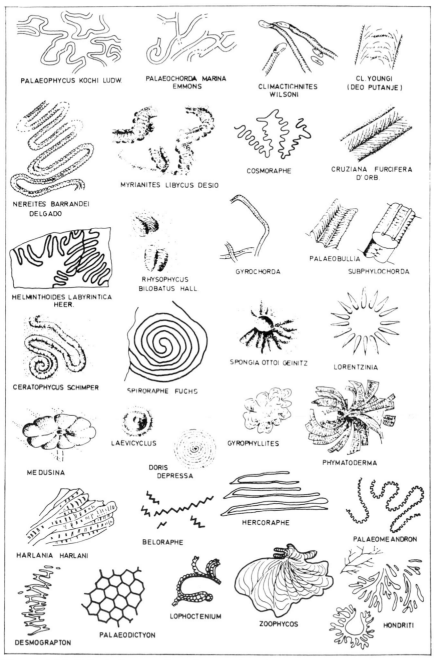

Figure 1.20 Selected examples of trace fossils as seen chiefly on bedding planes (Dimitrijević and others, 1968, Fig. 43).

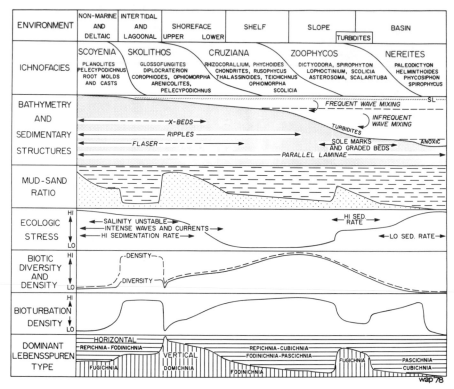

Figure 1.21 Suggested interrelationships between depositional environments and trace fossils (types and abundances).

rates, whereas the presence of escape structures (Fugichnia) and sparse trace fossils are indicative of high sedimentation rates. Suspension- or filter-feeding organisms usually construct vertical or U-shaped burrows (Domichnia) in unstable substrates, where bottom currents are active; however, where the substrate is stable and nutrient-rich and bottom currents are slight, deposit feeders usually construct shallow, horizontal feeding burrows (pascichnia or fodinichnia).

It is now possible to summarize the distribution and character of ichnofacies and their trace fossil assemblages in relation to bathymetry, sediment type, ecologic stress, and biotic diversity and density (Fig. 1.21). Muddy sediments clearly have the most abundant and diverse trace fossils and the highest degree of bioturbation. Basinal muds are usually intensely bioturbated with many grazing tracks, because slow sedimentation rates permit small numbers of organisms to completely rework the bottom muds, producing trace fossils of the *Nereites* assemblage. However, if anoxic conditions are present, most organisms are prevented from inhabiting the oxygen-deficient muddy environment and so very few tracks and trails are developed. Slope and turbidite muds have *Zoophycos* and *Nereites* trace fossil assemblages and the turbidite sands often have abundant escape structures (Fugichnia), with repichnial and pascichnial sole marks.

Low sedimentation rates and stable conditions in the shelf and lower shoreface environments promote high biotic diversity (many different species) and density. These muddy sediments have the greatest diversity of trace fossils and a high degree of bioturbation. The *Cruziana* assemblage commonly predominates in this environment. Fecal pellet muds, generated by filter-feeding and deposit-feeding organisms, are also abundant. Intense wave and current activity in the upper shoreface and beach zones results in unstable substrate conditions. Here, deep-burrowing, filter-feeding organisms predominate in a sparse and low-diversity biotic community and produce the vertical lebensspuren of the *Skolithus* ichnofacies—clearly one that is not common in shale facies. Muddy sediments are not abundant in the shoreface zone and, when present, are largely the result of fecal pellet accumulations.

Intertidal and lagoonal environments have high ecologic stress as the result of variable salinities, intense wave and current activities, and high sedimentation rates. However, muddy substrates and abundant food supply promote a high population density in a low-diversity biota and the result is a high degree of bioturbation. Fecal pellets of filter-feeding and deposit-feeding organisms are important components of these muddy sediments.

Nonmarine and deltaic environments are highly stressed, with high sedimentation rates, and consequently have low biotic diversity and density. Trace fossils and bioturbation are not abundant in these muds and sands. Plant root burrow systems are the most common type of trace fossil and bioturbation in the muds of the nonmarine and transitional environments.

Rhoads (1970, pp. 403–404) studied the mud substrate properties and burrow character of bivalves and concluded that the mass properties and stability of mud surfaces are important parameters in benthic ecology. Morphologic features of pelecypods and brachiopods, such as spines, small size, and low bulk density, indicate soft muds. He also concluded that cohesiveness and original water content may be inferred from the character of bioturbation. Sharp, well-defined burrows are indicative of cohesive muds, whereas indistinct or "fuzzy" bioturbation is indicative of high-water-content muds.

Fecal pellets are another type of biogenic structure. Many marine organisms feed by filtering suspended matter from sea water. Suspended argillaceous minerals may be ingested in this matter, pass through the digestive system, and be defecated as clay-rich pellets. The sizes, shapes, and constituents of the pellets are frequently unique to the specific organism that have made them (H.B. Moore and Kruse 1956). Argillaceous pellets produced in this manner are often sand or granule sized, with ovoid or cylindrical shapes. These pellets may be relatively firm and resistant to disintegration and are transported in the traction mode in the manner of sand grains. Pryor (1975) concluded that fecal pellets may be the single most important source of deposited muds in shallow marine, nondeltaic environments.

MINERALOGY AND GEOCHEMISTRY

More is known about the mineralogy and geochemistry of shales than anything else, because both have been extensively studied for many years. Accordingly we did not feel it necessary to review this literature at length. Instead, we offer a few comments on some of the most significant results, and how they might be applied to geologic problems related to shales.

Minerals in Shales

Shales contain a wide range of minerals (Table 1.6), although with very few exceptions only the clay minerals have been widely studied. There is, of course, good justification for this emphasis on the clay minerals because they can predominate (Table 1.7).

Clays have a micalike structure consisting of alternate layers of silica and alumina tetrahedra with, in most clay minerals, a layer of exchangeable cations (Fig. 1.22). A very complex mineralogy exists based on the number of sheets, the way the sheets are stacked (polymorphism), the substitution of such elements as Mg^{2+} and Fe^{2+} in the sheets, and on the nature of the exchangeable cations (Table 1.6). Methods of analysis are given by Carroll (1970).

The type of clay found in a shale is a function of provenance (rock type and climate) and diagenetic history. Depositional environment was once thought to exert a considerable influence on clay mineralogy through early mineral transformations in the basin of deposition, but it is now known that alteration of the clay framework does not occur, although there is a change in the exchangeable cation population (Russell 1970) that may have some usefulness as a paleosalinity indicator (Spears 1973). One possible *physical* effect of depositional environment on clay mineralogy is differential flocculation: kaolinite seems to flocculate at lower salinities than smectite–illite (Edzwald and O'Melia 1975) and so is deposited closer to the river mouth. Gibbs (1977), however, has argued that this results simply from the larger grain size of kaolinite.

In contrast, provenance is a well-documented control on clay mineralogy as first demonstrated by Biscaye (1965), in Atlantic Ocean sediments, where humid tropical terrains of moderate relief have produced kaolinite and gibbsite. As rainfall decreases, smectites become important (Sherman 1952). Illites probably derive from weathering of preexisting muscovites or illites, and chlorites from preexisting chorites; thus certain rock types generate particular mineralogies. Another well-known example of this effect is the

Table 1.6. Constituents of Shales: Origin and Significance

Framework silicates

Quartz: Forms 20%–30% of the average shale and is almost always present and is probably almost all detrital. May be, in part, eolian in origin, but very little is known about its occurrence, largely because it is too fine grained to study easily in thin section. However, the amount of quartz in a shale may be indicative of shoreline proximity. Other varieties of silica that may be present and should be looked for include chalcedony, opal CT, and amorphous silica, all of which may have originally been biologic (Marata and others, 1977)

Feldspar: Nearly always less abundant than quartz, and plagioclase is believed to be more abundant than potash feldspar. Part of the feldspar in a shale may be authigenic, but little is known on how to distinguish very fine detrital from authigenic feldspar in shales. It appears that even less is known about feldspar in shales than about quartz.

Zeolites: Commonly present as an alteration product of volcanic glass, but can also be found in the muds of hypersaline lakes (analcite). Modern marine sediments have phillipsite and clinoptilolite as the most common zeolites, where they may form a small percentage of the mud. Zeolites are useful indicators of very low-grade metamorphism in shales (laumontite, 50 to 300°C, and prehnite, as low as 90°C).

Clay minerals

Kaolinite (7 Å): Forms in soils developed under abundant rainfall, good drainage, and acid waters. Characteristic of tropical and subtropical weathering. In marine basins, concentrated near shore and therefore a good indicator of paleogeography in even the most ancient of basins. There are two other important varieties, dickite and halloysite, but little is known about their occurrence in shale. Another 7-Å clay mineral is chamosite, which is common in oolitic iron ores and may be overlooked in many shales.

Smectite–illite–muscovite (10 Å and greater): This is a structurally complex group, the members of which can form in different ways. Smectite, which is a hydrated expandable mineral, commonly forms from volcanic glass (bentonites) and also is common in many alkaline soils. It converts to illite during burial diagenesis via an intermediate mixed layer phase. Illite, by far the most abundant clay mineral, seems to be largely derived from preexisting shales and is also the principal clay mineral found in deeply buried shales, where it is associated with the chlorites. Muscovite is the end product of the diagenesis of illite, but also occurs as a detrital particle in unaltered shale, where it is commonly the coarsest clay mineral and is concentrated, along with biotite, on bedding and lamination surfaces. The thermal history of illites can be deciphered by identifying two structural types, 1M and 2M, and by measuring crystallinity. A special variety of illite–smectite is glauconite, an iron-rich variety. Glauconite seems to be exclusively marine and forms during slow sedimentation.

Table continued on pp. 48 and 49

Table 1.6. *Continued*

Chlorites, corrensite, and vermiculite: Chlorite is very sensitive to weathering and therefore rare in tropical and subtropical soils. Forms diagenetically with burial, especially in Mg-rich pore waters, and is commonly the second most abundant clay mineral in Paleozoic and older shales, forming as much as 70% but typically only about 10%–20%. During diagenesis, vermiculite appears to convert into corrensite and finally into chlorite.

Sepiolite and attapulgite: Both are Mg-rich clays that form under special conditions when pore waters are rich in Mg, such as in saline lakes. They have also been reported in small quantities in recent marine muds that are associated with volcanic activity.

Oxides and hydroxides

Iron oxides and hydroxides: Iron oxides or hydroxides are commonly present in shales, mostly as coatings on clay minerals. Such iron coatings are converted, in reducing environments, into pyrite or siderite. Hematite is the common iron oxide in shales, but in modern mud and weathered shales hydrous forms, such as goethite or limonite, are probably more common.

Gibbsite: The ultimate product of acid leaching, consists of $Al(OH)_3$. Characteristic of extreme tropical weathering, where it forms bauxites. May be associated with kaolinite in marine shale, the clay minerals of which have been derived from the weathering of a tropical landmass.

Carbonates

Calcite: Probably more common in marine than nonmarine shale, but there should be no carbonate minerals of any kind in marine mud deposited below the calcite compensation depth (unless introduced by turbidity currents). However, as with quartz and feldspar, very little is known about the distribution and form of calcite in shale.

Dolomite: Appears to be common in shale, but its relation to calcite is unknown. Like calcite, it may be an important cementing agent.

Siderite and ankerite: Common in concretions. In marine shales, they indicate intermediate Eh values and so may, when compared with pyrite occurrences, be helpful in paleogeographic reconstructions of those basins that have a more strongly reducing, deep part.

Sulfur minerals

Sulfates: Gypsum, $CaSO_4$ $2H_2O$, and anhydrite, $CaSO_4$, and barite, $BaSO_4$, occur as concretions in shale and may indicate some type of hypersalinity either during or after deposition. See Hounslow (1979) for a staining method to distinguish gypsum from anhydrite.

Sulfides: The only abundant sulfides in shales are those of iron. Modern muds contain several amorphous varieties, but shales have only crystalline FeS_2, as pyrite or marcasite. Both are much more abundant in marine shales than in continental ones and both indicate strongly reducing conditions either at the sediment–water interface or within the sediment. Pyrite and marcasite are indistinguishable in the field and therefore their relative abundance is not well known. There is some evidence that marcasite forms under lower pH conditions than pyrite.

Table 1.6. *Continued*

Other constituents

Apatite: A phosphatic mineral that forms, when surface waters have much organic productivity. Commonly apatite forms nodules in slowly deposited marine muds. Another phosphate mineral is vivianite, an iron-bearing phosphate, which may well be more abundant than we think.

Glass: Common in modern muds of either continental or marine origin that are associated with volcanism. Probably converted during burial to either zeolites or smectites and free silica.

Heavy minerals: Potentially legion and possibly best preserved in concretions, although little is known about their occurrence and abundance in shale.

Organic materials

Discrete and structured organic particles: These are mostly either palynomorphs or small coaly fragments (vitrinite). Very useful in Phanerozoic shales for correlation and can be used for helping to identify proximity to shorelines. Reflectance and color help define the thermal history of the basin.

Kerogen: Amorphous organic material that systematically changes color with increasing temperature and finally converts to graphites. Present in almost all shales except red ones and tells about the gas and oil potential of a basin and its thermal history. Chemical characterization is complex.

common association of smectites with volcanic terrains. Also, low-silica parent rocks, such as syenites, favor the development of gibbsite.

The usefulness of this provenance pattern is somewhat obscured, however, by later diagenesis. It has long been noted that Paleozoic shales tend to be almost entirely chlorite + illite (Weaver 1967), while younger rocks have, in addition, abundant kaolinite and smectite. F.T. MacKenzie (1975) has argued that such changes reflect increasing diagenesis. Certainly the transition smectite $\rightarrow$ illite is common with burial to more than 10 000 ft (Aronson and Hower 1976), and probably accounts for the scarcity of smectite in older rocks. However, Weaver (1967) felt that the appearance of kaolinite in younger rocks was related to development of land plants with a concomitant increase in the intensity of chemical weathering.

Table 1.7. Average Clay Mineralogy (Pettijohn 1975, Table 8-3)

	Percentage
Clay minerals	58
Quartz	28
Feldspar	6
Carbonates	5
Iron oxide	2

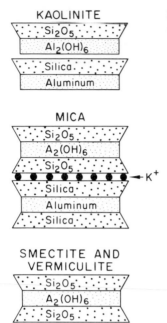

Figure 1.22 Generalized structure of major clay mineral groups.

Shales contain a great variety of nonclay minerals, which make up a significant fraction of these rocks (Table 1.6), but these have not been sufficiently studied. For instance, what is the average albite/K-feldspar ratio, and how does it relate to provenance and diagenesis? Do carbonates occur as cement or discrete particles? What kinds of carbonate minerals exist in shales and how, if at all, do they vary with either depositional environment or geologic age? How many phases of silica are commonly present and what controls them? To our knowledge none of these questions has been addressed.

Inorganic Geochemistry

Inorganic geochemical studies of shales have mostly been of trace elements and most commonly have tried to assign the various elements to particular minerals or organic substances within the rock rather than to determine provenance or depositional history. An exception is the attempt, largely unsuccessful, to relate trace elements to paleosalinity (Cody 1971). Major elements deserve more attention because their study, when combined with careful mineralogy and petrology, should be useful for interpreting the histo-

Table 1.8. Chemical Composition of Average Shale and Major Sandstones Types (Data from Pettijohn 1975, Tables 7-3 and 8-7)

	Average shale (%)	Ortho-quartzite (%)	Graywacke (%)	Arkose (%)
SiO_2	58.10	95.4	66.7	77.1
Al_2O_3	15.40	1.1	13.5	8.7
Fe_2O_3	4.02	0.4	1.6	1.5
FeO	2.45	0.2	3.5	0.7
MgO	2.44	0.1	2.1	0.5
CaO	3.11	1.6	2.5	2.7
Na_2O	1.30	0.1	2.9	1.5
K_2O	3.24	0.2	2.0	2.8
CO_2	2.63	1.1	1.2	2.8
C	.80	—	0.1	—
H_2O	5.00	0.3	3.0	0.9

ry of a basin. When this is done, the analyses must be related to a carefully detailed internal stratigraphy of the shale. Unfortunately there are as yet no accepted models, comparable to those for sandstones, for relating composition to basin development. For example, what is the analog of a graywacke or an arkose in shales?

To illustrate this point, compare the distribution of elements in shale with that in various types of sandstone (Table 1.8). Instead of being mostly SiO_2, shales have substantial amounts of other components and therefore should provide ample opportunity for classification schemes based on composition. One possibility is to divide shales into suites according to their associated sandstone types, as originally suggested by Krynine (1948) and discussed by Weaver (1978, pp. 160–161). Another possibility is to use the petrology of the shales themselves (optical or X-ray) to establish categories, such as feldspathic or calcareous, to which composition can be related. A scheme for relating composition to provenance has been proposed by Björlykke (1974, pp. 262–263), who showed that the ratio $(Al_2O_3 + K_2O)/(MgO + Na_2O)$ could be used to detect the presence of a volcanic-arc provenance in some Ordovician shales in Norway. Finally, Englund and Jorgensen (1973) have also proposed a chemical classification system for shales that may be useful in provenance studies.

Although provenance probably dictates most of a shale's composition, the environment of deposition does have some effect. For instance, Murray (1954) showed that marine clays in the Pennsylvanian of the United States have considerably more calcium than do nonmarine clays. Also, Spears (1973) has suggested that the distribution of exchangeable cations, particularly the ratio Mg^{2+}/Ca^{2+}, can be used to infer paleosalinity.

Exchangeable cations differ from the other major elements in shale because they can easily be displaced from clay minerals by other ions in solution. They are therefore chemically much more reactive than other elements in a shale. They should always be separately determined in geochemical analyses, because they are controlled by pore-water chemistry rather than by provenance. Postdepositional processes, it should be noted, seem to have little effect on bulk composition (Shaw 1956).

Trace element geochemistry has been exhaustively studied, as shown by the references in the Annotated and Illustrated Bibliography (Chapt. 3). However, little success has been achieved in attempts to relate trace elements to depositional environments. Perhaps greater utility will be found in using them for stratigraphic correlation, such as in Fenner and Hagner's (1967) study of the Esopus Formation. Gamma-ray logs are a type of trace element tool. Most gamma radiation is generated by uranium, which is controlled by the amount of organic matter, an important stratigraphic variable. Potassium and thorium also generate gamma rays and wire-line tools are now available that can discriminate among these sources. Thus it is now possible to measure three elements of possible stratigraphic utility in bore holes or in outcrop.

Geochemical properties of shales change with time. For example, several workers have reported significantly higher K_2O in early Paleozoic shales than in younger shales (Fig. 1.23). At present, we do not know whether such changes reflect increased diagenesis in the older rocks or a change in the chemistry of the earth's surface, perhaps related to the appearance of land plants (Weaver 1967, p. 2187).

Organic Geochemistry

Our understanding of the organic chemistry of shales has seen great progress in recent years, largely as a result of interest in shales as source rocks for petroleum (Tissot and Welte 1978). The capacity of a shale to generate hydrocarbons is governed by the amount of organic matter present, its type, and its state of thermal maturity. The minimum amount of organic matter for the generation of significant amounts of hydrocarbons seems to be about 0.5% organic carbon, with known source beds averaging 2.2% (Tissot and Welte 1978, p. 430). The type of organic matter, woody or amorphous, determines whether only gas or oil and gas will form (Fig. 1.24). Thermal history also affects the type of hydrocarbon produced: Note that non-biogenic methane requires higher temperatures than oil. In addition, Fig. 1.24 shows that source beds pass through an optimum temperature during diagenesis, beyond which the hydrocarbons are destroyed. Consequently, the determination of the temperature history of shales is an extremely important part of basin analysis. Some of the methods that have been used are listed in the Question Set (Chapt. 2). Time is possibly important too (Connan 1974): The same degree of diagenesis of the organic matter can be achieved at lower temperatures if the time is longer.

The organic matter in shales can also be used as a type of fossil. For instance, several indicators are available for detecting nonmarine carbon de-

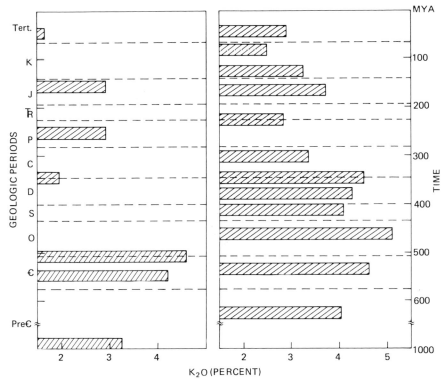

Figure 1.23 Average K_2O content of shales in geologic systems: Australian data at left (van Moort, 1972, Table 1) and Russian data at right (Vinogradov and Ronov, 1956, Fig. 2).

posited in marine shales. The best is probably $^{13}C/^{12}C$ ratio (Hedges and Parker 1976); another possibility is the presence of lignin derivatives (Gardner and Menzel 1974). It should be possible to use these to detect proximity to the shoreline in an ancient basin. To our knowledge, however, no one has been able to use biochemical fossils as conventional time-stratigraphic markers. They have been sought in Precambrian rocks as indicators of early evolution, but contamination seems to cause great difficulties (Leventhal and others 1975).

COLOR

The color of sedimentary rocks, and of shales in particular, has long been a subject of controversy among geologists. Unfortunately, very little progress has been made either in understanding the origin of color or in using it to solve geologic problems. Color is the most obvious feature of a shale. It is valuable for stratigraphic correlation and seems to have possible environmental significance. For example, red shales commonly occur

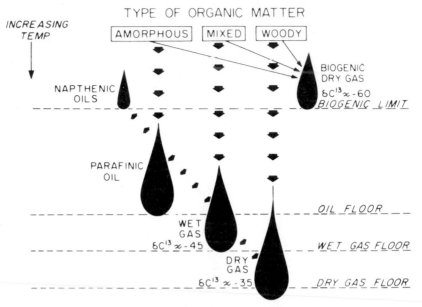

Figure 1.24 Generation of hydrocarbons from different types of organic matter. Increasing temperature leads to the conversion of all organic matter into gas. (Redrawn from Dow, 1978, Fig. 2, with additional data from Phillippi, 1977, p. 43, and Stahl, 1975.)

continental environments, black shales with restricted marine basins. Unfortunately, a number of other environments can produce these colors. Deep-sea clays are commonly reddish brown in the modern oceans and apparently in some ancient rocks as well (Ziegler and McKerrow 1975). Similarly, estuarine or tidal flat deposits often have black mud accumulations. Can we find an explanation for this diversity?

For all its interest, very few workers have investigated the source of color in their rocks. An exception is the voluminous literature on red beds reviewed by Van Houten (1973). The essential conclusion of this work seems to be that color in sediments, because it can be changed so easily, is almost always of depositional or diagenetic origin rather than detrital. In other words, red sediments are produced by oxidizing depositional environments, not by red soils (Berner 1971, p. 197).

An important early study of shale color was made by Tomlinson (1916) and amplified by Pettijohn (1975, Fig. 8-9). Tomlinson showed that the color of slates is independent of the total amount of iron present but strongly controlled by the Fe^{3+}/Fe^{2+} ratio. High ratios are associated with red colors, low with greens. This idea has been confirmed by McBride (1974) in a study of continental shales. He found a progression—red $\rightarrow$ yellow $\rightarrow$ green $\rightarrow$ gray—corresponding to decreasing Fe^{3+}/Fe^{2+}, but with fairly constant total iron. A number of other workers, however, have

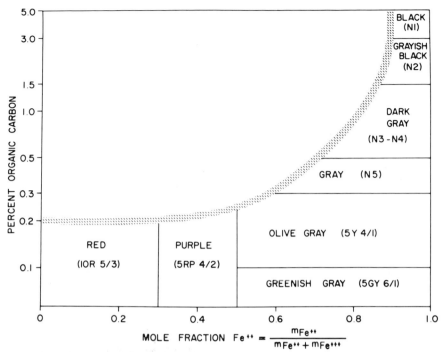

Figure 1.25 Suggested relationship of shale color to carbon content and oxidation state of iron. The mole fraction is used to indicate the proportion of the total iron that is in the +2 state and m represents the number of moles of iron per gram of rock. Finer subdivisions of color are possible, but are difficult to reproduce. Colors determined on wet samples in natural light.

documented cases of nonmarine strata deposited under arid conditions, in which green or gray interbeds or blebs in predominantly red sequences have significantly less iron than the associated red beds, presumably because they once contained Fe^{3+} that was removed after being reduced to the more soluble Fe^{2+} (Picard 1965, Friend 1966, p. 282, T.R. Walker 1967, p. 361). Such removal probably only occurs when the rocks are deposited above the water table. The data of McBride (1972) and Tomlinson (1916) show, however, that removal of *iron* is not necessary to develop green colors, only *reduction* of the Fe^{3+} to Fe^{2+}.

We suggest that the amount of organic carbon present is another important, and partly independent, control of color (Fig. 1.25). This diagram was constructed using data on Fe^{3+}/Fe^{2+} from Tomlinson (1916), Fe^{3+}/Fe^{2+} and C from McBride (1972), and approximately 100 organic carbon measurements made in our laboratory. The Geological Society of America rock color chart (Goddard and others 1975) was used for the color designations. There are several important qualifications to be kept in mind when using this diagram.

The boundaries are based on averages; consequently, individual samples may fall into the wrong fields. Secondly, different operators seem to consistently choose darker or lighter colors in the NI–N6 series, thereby shifting the scale. As a result, the fields shown in Fig. 1.25 are not precise, although they still reflect the relative positions of the various colors with respect to Fe^{3+} and C. Trask (1937, Fig. 2) also reported a relationship between carbon content and color in sediments that showed a consistent trend, but with considerable spread. Berner (1971, p. 207) has also shown that the metastable iron sulfides (FeS) can impart a strong black color to modern sediments with a modest carbon content ($< 1\%$). In ancient rocks, however, these minerals have all been converted to pyrite (FeS_2), which is not an important pigment. Additional factors that influence color are grain size, mineralogy, and thermal maturity of the organic matter, but these seem to be relatively minor.

Notice that the samples fall along two sides of the diagram. This distribution is caused by the effect of organic carbon on Fe^{3+}/Fe^{2+}; even small amounts of organic matter favor the reduced form (Fe^{2+}). Once there is enough carbon present to be visible, the iron is already converted to Fe^{2+}. Shales seem to partition into two series, a red $\rightarrow$ purple $\rightarrow$ greenish gray series based on Fe^{3+}/Fe^{2+} and a greenish gray $\rightarrow$ gray $\rightarrow$ black series based on carbon content. In effect, pigmentation is by Fe^{3+} in the first series, but by carbon in the second, once the Fe^{3+} has been converted to Fe^{2+}. Light colors in the 7 and 8 values seem to be rare in unweathered shales. They are common in soils and underclays and so may occur when most of the iron has been leached away and neither Fe^{3+} nor carbon remains as a pigment. Blueish colors, such as pale blue (5B6/2) or pale blue green (5BG7/2), were not encountered in our study but are common in calcareous shales and so may be a sensitive indicator of the presence of carbonate minerals.

Shale color, then, is controlled by two "rock" variables that are directly measurable, Fe^{3+} and carbon. In order to relate color to depositional environment, we must know how these rock variables are affected by environmental variables. Because the Fe^{3+}/Fe^{2+} ratio is controlled by the oxidation state, which is in turn controlled by the amount of organic matter in sediments, all color in shales is ultimately controlled by the amount of organic matter present. We propose that three environmental variables are important in controlling the amount of organic matter: the rate of production of organic matter in surface waters of the basin, or in some cases its introduction by rivers; the rate of sedimentation of other components, such as terrigenous particles or the shells of pelagic organisms, which serve to "dilute" the organic matter; and the rate of decomposition of the organic matter in the upper few centimeters of sediment. This rate is controlled by the amount of oxygen in the bottom water. That is

$$C = f(P, O, S)$$

with C the percent organic carbon, P the flux of organic carbon to the sediment in milligrams of carbon per square centimeter per year, S the sedimentation rate in milligrams of sediment per square centimeter per year, and O the rate of oxidation of the organic matter in the sediment in milligrams of carbon per square centimeter per year. One possible functional relationship among these variables is

$$C = (P - O)/S$$

which assumes that the only effect of sedimentation rate is to dilute the organic matter. For such relationships, the S term predominates if P and O are relatively constant. Curtis (1977) has in effect argued that this is usually the case in marine basins. However, based on our experience with shales of the Appalachian Basin, oxygenation of the bottom water is also an important control of color. For instance, interbedded black and gray shales are distinguishable by a universal lack of bioturbation in the black shales, suggesting anoxic conditions, whereas burrowing is common in the gray and green interbeds, indicating oxygenated bottom water (Byers 1977, p. 9). Here P and S are essentially the same so that O is the determining variable. In the same way, very high values of O, such as in arid fluvial environments, lead to oxidation of all of the carbon, producing red shales. Also note that if only sedimentation rate (S) were important, very slowly deposited muds such as those in the deep sea would be organic-rich. Instead, because of high O values, they are very low in organic matter. Reddish marine shales on continents may be explainable in the same way (Ziegler and McKerrow, 1975). Evidence that sedimentation rate has an effect beyond simple dilution has been presented by Toth and Lerman (1977) and Berner (1978). He studied the rate of reduction of sulfate by bacteria in muds that are being deposited at rates varying from 0.002 to 5 centimeters per year. The rate of bacterial activity, surprisingly, increases with increasing S:

$$k = 0.04 \ S^2$$

where k is a rate constant for the bacterial decomposition of organic matter.

The final step of relating these environmental variables to actual depositional environments needs to be tested by accumulating data on carbon content, sedimentation rate, and oxygenation from modern environments and comparing it to mud color. Several authors have reported colors from recent depositional environments (Rusnak 1960, Pelletier and others 1968) that seem to agree with our conclusions, but studies combining carbon content, color, and sedimentation rate are lacking.

Finally, we point out that shale color has an even greater utility in stratigraphy than in environmental analysis. In shale sequences, colors are nearly always the best guide to stratigraphic subdivision. Black and gray shale sequences are good examples, for here color contrasts correlate closely with the stratigraphy seen on gamma ray logs because the organic matter is in some way associated with uranium. And, of course, other shale colors are just as useful for stratigraphic subdivision even though they are harder to relate to wire line logs. Mapping lateral color variations of shales should be very useful to the study of their depositional environments.

There is still another aspect of color to consider—how, within a stratigraphic unit, does it vary across a basin? Studies that document the lateral variation of color are as yet rare and of special interest would be those that map it from shelf to deep basin or down dip on a continental margin. As color is traced down dip, how could one sort out the effects of differential compaction and diagenesis from those of depositional environment?

COMPACTION

Shales undergo important changes in physical structure brought about mostly by compaction, in addition to the diagenetic changes discussed under geochemistry and mineralogy. Such compaction is important because in thick shale sequences it causes significant tectonic movements, which produce shale diapirs (Fig. 1.26) and mud-cored anticlines. In this process it also expels large amounts of water. Deformed delicate fossils (Ferguson 1963), differential compaction around early formed nodules and/or concretions, and contorted dikes (Borradaile 1977) are excellent direct evidence of compaction (Fig. 1.27).

There is a great deal of current research on compaction and ideas are changing rapidly. Most work has centered on the problem of "overpressur-

Figure 1.26 Shale diapir cuts country rock (subhorizontal sandstones and shales) of Cretaceous age in Recôncavo Basin, north of Salvador, Bahia, Brazil. (Photograph courtesy of Petrobras.)

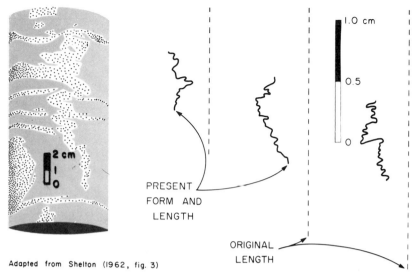

Adapted from Shelton (1962, fig. 3)

Figure 1.27 Present form (solid contorted line) and original length (broken line) of small dikes in Cretaceous shale. (Redrawn from Shelton, 1962, Fig. 3.) With this method the original thickness of shale interbeds was estimated to be 2.6 times greater than present thickness.

ing." Clays do not release water in a simple fashion during compaction. Instead, there is a tendency, more exaggerated in thicker shales, for water expulsion to lag behind burial, leading to a "compaction disequilibrium" (Magara 1975). During deeper burial this excess water pressure is often suddenly released, producing significant faulting that can form hydrocarbon traps (Bruce 1973). This type of structure can be important because it forms early, before hydrocarbons have been lost from the system. Also, because loss of water is retarded in these shales, abundant fluids are available to drive hydrocarbon migration in the depth range at which oil and gas are generated. Besides this simple undercompaction, high fluid pressures in shales can be generated by the expansion of water with increasing temperature (Barker 1972), by the generation of methane (Hedberg 1974), or by the dehydration of smectite (Powers 1967). This last mechanism may be particularly important for hydrocarbon migration in thinner shale sequences where undercompaction is not well developed or where structural traps form later, as in the Mowry Shale of Wyoming (Nixon 1973).

Geochemistry of the fluids may play a part in these processes, but as yet little work has been done. Most deeply buried shale waters are highly saline, perhaps because the shales act as semipermeable membranes, allowing water to pass through but retarding passage of charged species (Hanshaw and Coplen 1973). Conversely, shales may have lower salinity water than associated sands (Schmidt 1973), perhaps because the negative charges on the clay surfaces in well-compacted clays exclude chorine ions, pushing them into interbedded sands (Berner 1971, pp. 108–111).

OCCURRENCE AND CHARACTERISTICS OF MUD IN MODERN ENVIRONMENTS

Understanding the origin of modern muds is fundamental to improved interpretation of ancient shales. We have summarized the characteristics of mud in all the major environments, as far as our present knowledge permits (Table 1.9). Our table was inspired by and broadly follows similar attempts for both sandstones (W.S. MacKenzie, 1972, Table 2) and carbonates (Wilson 1975, Fig. 11-4) and was compiled largely from the study of modern sediments using both the literature and our own field experience. A few of the many references that were consulted are listed at the end of the table.

The 11 major sedimentary environments are all represented: five nonmarine environments (Table 1.9, Part A), three transitional environments (Table 1.9, Part B), and three marine environments (Table 1.9, Part C). Six characteristics were selected to help the reader better describe, recognize, and interpret both modern muds and their ancient counterparts. Geometry, areal extent, thickness, and abundance define the overall stratigraphic character of any deposit, especially from the viewpoint of the subsurface geologist. Generally, such characteristics are much better known for ancient shales than for modern muds. Areal mapping in either the outcrop or subsurface is essential to determine these properties, and more work of this type is needed in both modern and ancient sediments. The next three columns, lithologic character, bedding and sedimentary structures, plus associated lithologies, are all described using outcrops, cores, or cuttings and all are fundamental to the understanding of every sediment. Finally, the last column, process, attempts to explain how the mud has been transported and deposited.

What generalizations can be made about each of the major environmental groupings? Nonmarine muds are the most variable of all. For example, although many are very silty, some of those deposited on the ponded lakes of floodplains, in oxbows, and in large lakes can have very little silt and be exceedingly pure. Virtually every color can be found in nonmarine muds—black, gray, brown, red, and even white. Fauna are very restricted, but plant material can be very abundant, especially in wet climates. Finally, nonmarine muds are rarely widespread—except loess, which can cover thousands of square kilometers.

Marine muds, in contrast, are the most uniform, especially with respect to texture, and are the most likely to be fossil rich in every geologic period. Their more common colors are typically greenish gray, gray, dark gray, and black, although red colors are possible. Two notable features are the wide lateral extent of even very thin marine shales, an extent that far surpasses that of loess, and the great thicknesses of individual stratigraphic units in rapidly subsiding basins. Another notable feature is their high organic or

Table 1.9. Characteristics of Shale Deposits: Environmental Matrix
Part A. Nonmarine Environments

Environment	Geometry, areal extent, thickness and abundance	Lithologic character	Bedding and structures	Associated lithologies	Flora and fauna	Setting and process
Eolian Continental deserts	Small- to medium-sized sheets up to a few kilometers; and some small elongate patches; rare and mostly less than 2 m	Commonly silty; may be calcareous and gypsiferous; red, brown, yellow, tan, gray, black	Mud and dessication chips, ripples, horizontal lamination, and some deformation from crystal growth; minor bioturbation	Sand, silt, and some gravel with evaporites and algal deposits in saline lakes	Largely absent; but some vertebrates in Mesozoic and younger deposits	Settling and evaporation in ephemeral lakes and rare overbank deposits from wadis
Loess	Widespread sheets to thousands of square kilometers may be 10s of meters thick near source but thin exponentially downwind	Commonly well-sorted calcareous and noncalcareous silt that becomes finer downwind; tan and yellow	Lamination near source as well as some calcareous concretions and vertical jointing	Glacial deposits as well as buried bedrock of all types; generally contains buried soils	Chiefly land snails, rootlets, spores, and pollen	Differential settling of wind-borne clays and silts from the deflation of braided outwash streams as well as deflation of continental deserts
Alluvial Piedmont	Small and rare elongate patches trending downdip; generally less than 1 m	Poorly sorted silty pebbly mud; red, brown, yellow, tan, white, gray, and black	Planar lamination in association with silty ripple bedding as mud-flow deposits and some clay chips	Gravel, sand, and silt plus mud flows	Very rare, but some vertebrates in Mesozoic and younger deposits and some wood	Settling and evaporation of overbank muds, especially in distal part of alluvial fans; mud-flow deposition proximal
Floodplain	Sheets parallel to shoreline for 10s of kilometers, but can be interrupted by channels; mud may be dominant with thicknesses of a few to 10 or more meters	Very fine to silty and may be organic rich, especially, in humid climates; black, gray, brown, tan, red	Planar, thin lamination with minor deformed bedding. Also some ripples n silty muds near channe s and chips, clay chip conglomerates, and dessication cracks.	Sands and silts, plus peats and coals in humid climates as well as evaporites in arid climates	Logs and rootlets common as well as some Mollusca and vertebrate debris in Mesozoic and younger deposits	Settling and flocculation in overbank areas and abandoned meanders; proportion of mud generally increases with decrease in gradient
Lacustrine	Ovate to elongate valley fills ranging up to thousands of kilometers or more; mud predominates with thicknesses of 100s of meters or more	Fine to silty mucs that can be gypsiferous in arid regions as well as calcareous; black, gray, tan, red, yellow, and brown	Horizontal and wavy laminat on, some ripples in coarser s ltier muds, bioturbation and possibly dessica ion structures.	Silt, sand, and minor gravel plus sapropels as well as evaporites and carbonates	Ostracods and gastropods—invertebrates as well as plant debris, spores, and algae	Settling and flocculation in tectonic basins as well as in glacially scoured and damned depressions plus pcnding in tributaries of alluviating rivers

Table 1.9. Part B. Transitional Environments

Evironment		Geometry, areal extent, thickness and abundance	Lithologic character	Bedding and structures	Associated lithologies	Flora and fauna	Process
Deltas Wave dominated		Scattered, small, thin elongate patches on shore with elongate to ovate sheets parallel to coast on shelf and slope where mud may be up to 100s of meters thick	Mostly fine, but silty near distributary mouths; gray, black, and tan	Thin laminations becoming coarser proximal to mouth; abundant bioturbation	Mostly sand and minor gravel onshore, but silts and fine sands and possibly even some minor carbonates offshore	Minor lacustrine fauna onshore with open marine offshore; low-diversity, transitional fauna; Scoyenia trace fossil assemblage	Flocculation very important; most of mud deposited as a wedge offshore from jet-suspended clays, downcurrent from mouth
River dominated		Very abundant mud, mostly as sheets with some clay plugs from abandoned distributaries; thicknesses to 10s and 100s of meters	Mostly fine muds, many very organic rich; gray, black, brown, and tan	Lenticular plus well-laminated muds, rootlets common onshore; some minor interbeds of sand; bioturbation	Sands and silts on splays plus peats and coals in humid climates and even carbonates; evaporites possible in arid climates	Plant rootlets onshore plus logs and wood; open marine offshore; fairly low-diversity fauna; Scoyenia trace fossil assemblage	Overbank deposition onshore, but with flocculation offshore as above
Estuarine Low wave energy		Dissected sheets of moderate to small size plus mud banks a few 100 to 1000s of square meters; abundant mud with thicknesses up to about 10 m	Dominantly fine mud with laminated silts, which may be organic rich; gray, black, and brown	Interstratified muds, silts, and sand (flaser bedding); some of latter with bipolar cross laminations; possible storm layers; intense bioturbation and some pelletal muds	Mostly silts, but some sands	Brackish to restricted faunas, commonly with low diversity, but fairly abundant	Flocculation, differential settling, and biogenic pelletization in drowned estuaries along coasts with low tidal range and weak onshore wave climate
High wave energy		Minor mud as small- to medium-sized patches a few meters thick	Coarse to medium muds with much silt; gray, black, and brown.	As above, but probably with less bioturbation	Chiefly sands and silts, but gravel also possible	Brackish to open marine, with some allochthonous fauna	Flocculation and biogenic pelletization in bays along coasts with high tidal ranges and/or a strong onshore wave climate
Tidal		Sheets parallel to main channels plus cresent-shaped clay drapes on point bars; commonly very abundant and can exceed 20–30 m	Fine to silty organic-rich muds; gray, black, and brown	Laminated and well-burrowed muds interlaminated with rippled silts and sands (flaser bedding), clay galls, and shell pavements, plus dessication structures on intra- and supratidal portions; abundant bioturbation	Mostly terrigenous silts and sands, but also carbonates	Abundant, low-diversity marine fauna; Skolithus trace fossil assemblage	Flocculation, biogenic pelletization, and entrapment by algal mats and plant baffles; mud tends to be concentrated inshore by unequal tidal currents, in all except coasts with very high tidal ranges

Table 1.9. Part C. Marine Environments

Environment	Geometry, areal extent, thickness and abundance	Lithologic character	Bedding and structures	Associated lithologies	Flora and fauna	Process
Marine Shelf Inner	Elongate sheets many kilometers long to small patches; rare to abundant mud a few to 10s of meters thick	Fine to silty muds some of which contain fecal pellets; gray, green, brown, and black	Graded storm bedding in associated sands and silts; some flaser bedding with ripples mostly parallel to shore; much bioturbation with feeding trails predominant	Some sands and silts, some of which may be relict; possibly carbonate seaward	Open marine with high diversity; *Skolithus* and *Cruziana* trace fossil assemblages	Deposition from suspension and biogenic pelletization along protected, low-energy coasts, but also on open coasts, when mud supply is very great; wave energy and supply control mud abundance
Marine Shelf Outer	Wide sheets to rare patches; rare to abundant mud; a few to 10s of meters thick	Fine to silty muds some of which are fecal pellets; gray, green, brown, and black	Horizontal lamination and minor ripples; rare slumping; mixed feeding structures below wave base	Silts and sands shoreward, some of which may be relict; some carbonate possible	Open marine with high diversity; *Cruziana* trace fossil assemblage	Deposition from suspension and biogenic pelletization below wave base; if mud supply is limited, carbonate deposition; wave energy and mud supply control mud abundance
Slope and Rise	Wide sheets commonly interrupted by narrow downdip-trending channels, some of which are mud filled; mud common with thicknesses to 100s of meters and more	Silty muds finely interlaminated with pelagic muds; gray and black	Graded and massive bedding plus slumps and slides and clay-pebble conglomerates in channels; some escape structures; commonly trace fossil zonation is closely related to depth	Sands, silts, and muds as well as some carbonates	Mixed deep and shallow benthic fauna; *Zoophycus* and *Nereites* are most common trace fossil assemblages	Resedimented gravity and suspension flows, slumping, and some pelagic deposition with some contour currents possible on lower slope; wave energy and mud supply control mud abundance
Deep Sea	Wide sheets to 100s of kilometers interrupted by rare channels of silt and sand; mud predominates with thicknesses of 10s to 100s of meters	Normally the finest and most silt-sand free of all muds; also carbonate free, if below compensation depth; often rich in organic matter; gray and black	Some thin laminae of silt, if not too much bioturbation	Fine sands of distal turbidites near continental rise as well as pelagic oozes and manganese nodules	Calcareous planktonic assemblages above compensation depth and siliceous below plus some windblown phytoclasts. Trace fossils mostly the *Nereites* assemblage	Some suspension flows, but mostly pelagic deposition and pelletization by zooplankton with some contributions from eolian dust and ash falls; deep oceanic currents may help disperse mud

Table 1.9. Part D. Selected References

Continental deserts	Estuarine
Glennie (1970)	Folger (1972)
Groat (1972)	Klein (1967)
McGinnies and others (1968)	Mayou and Howard (1975)
Loess	Oomkens and Terwindt (1960)
Flint (1971)	Tidal
Simonson and Hutton (1954)	Evans (1975)
Waggoner and Bingham (1961)	Reineck and Wunderlich (1968)
Alluvial	Van Straaten (1961)
Piedmont	Deltas
Blissenbach (954)	Coleman and Wright (1975)
Groat (1972)	Fisk and others (1954)
Williams (1973)	Scott and Fisher (1969)
Floodplain	Scruton (1960)
Fisher and Brown (1972)	Shelf
Fisk (1947)	Gibbs (1973)
Lacustrine	Howard and others (1973)
Hutchinson (1957)	McCave (1972)
Picard and High (1972)	Reineck and Singh (1966)
Reeves (1968)	Sutton and others (1970)
	Swift (1969)

kerogen content—only marine muds are significant generators of petroleum.

The transitional deltaic, estuarine, and tidal environments fall between these two extremes, although estuarine and tidal muds are closer to those of the marine environment and appear to be much less variable than muds of the detaic environment. For example, muds of deltas can occur as restricted deposits in abandoned distributary channels or be more widespread in adjacent interdistributary swamps or in freshwater lakes in the delta plain, and they can occur as very widespread prodelta deposits, which volumetrically are generally the most important. Marine fauna, when present in the muds of transitional environments, has a low diversity but may be very abundant. Associated lithologies can also be very diverse depending upon the quantity of mud brought to the depositional basin and upon climate. Finally, mud is always more abundant in the low-energy deltas and estuaries than in their high-energy equivalents.

PALEOCURRENTS IN SHALY BASINS

Two approaches are available to determine the paleocurrent system of shaly basins: to study paleocurrent structures in interbedded siltstones, sandstones, and skeletal carbonates, or to study directional or scalar properties of the shale itself. Normally, the former is easier and quicker. Siltstone, sandstone, and/or carbonate are often present in shaly basins, so as a first approach this is the best path to follow. Finely textured sole marks (Fig. 1.28) and oriented clasts (Fig. 1.29) can be present on the thinnest of interbedded siltstones.

The shales are more difficult to study, in part because their depositing currents were weak. However, suppose only shales are present? If so, the tools

Figure 1.28 Top: Finely textured sole marks on base of thin, very-fine siltstone interbedded with Devonian shale sequence of Appalachian Basin. With oriented cores it is easy to measure and record paleocurrent indicators such as these. Bottom: Carbonized wood—measure azimuth of boundary to estimate paleocurrent.

available include: (1) measurement of fossil orientation in the shale itself; (2) measurement of the shape orientation of virtually ever-present silt in the shale as well as clast orientation; (3) mapping scalar properties to define dispersal patterns that can be related to basin shape; and (4) measurement of the orientation of concretions.

In pure shales, fossil orientation has been the most widely used paleocurrent indicator; studies of graptolites in Lower Paleozoic shales are especially notable. Because they were light and elongate, graptolite rhabdosomes responded to the most gentle of currents and, as a result, provide sensitive indicators of paleocurrent systems (Fig. 1.30). Most studies (cf. Moors 1969) show remarkable little variance in their orientation, Ruedemann's (1897) study in the Utica Shale being a good example (Fig. 1.31). To judge the importance of the paleocurrent systems in muds and shales, simply ask yourself, after examining Ruedemann's map, how complete any interpretation of the Utica Shale would be if it did not consider the uniform orientation of the graptolite rhabdosomes that Ruedemann measured? Fine charcoal and

Figure 1.29 Oriented clasts at base of thin siltstone from Devonian shale sequence of Appalachian Basin. Long axes parallel current.

Figure 1.30 Graptolite orientation in a Silurian shale from Poland (Jaworwski, 1971, Pl. 36, Fig. 3).

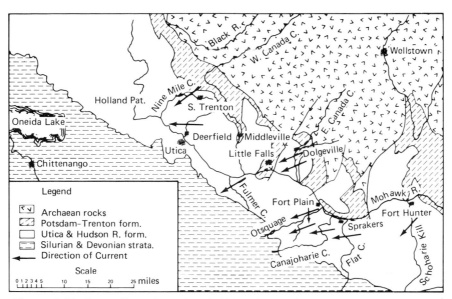

Figure 1.31 Graptolite orientation in Utica Shale (Ordovician) of New York as measured by Ruedemann (1897). His was the first paleocurrent map to show measurements at outcrop.

woody debris also promise much potential, judging by the few published studies (Fig. 1.32) and our studies of Devonian shales in the Appalachian Basin. Few studies have been made in other shales, even though charcoal orientation is commonly reported in the siltstones and sandstones of turbidites, where it is generally well oriented and correlates well with the primary directional structures such as sole marks and parting lineation. Not much is known about the transport processes and transport paths of woody plant debris as it is dispersed into marine basins, although our experience suggests that bimodal orientations may be the rule for elongate woody fragments.

Oriented brachiopod valves (especially *Lingula*), high-spired gastropods, ostracodes, and forams (Foraminifera) have all been seen in mudstones and shales and can have remarkably little variance in orientation in single outcrops. An interesting example of this is the good orientation of ostracods in the mudrocks and shales of the Pennsylvanian Monongahela and Dunkard strata of West Virginia reported by M.L. Jones and Clendening (1968). To this list should be added cephalopods and other conical forms, such as *Tentaculites*. M.L. Jones and Dennison (1970) have made one of the most comprehensive studies of fossil orientation in the Paleozoic shales of the Appalachian Basin. They measured more than 13 000 fossils in shales ranging in age from Middle Ordovician to Late Devonian and found preferred, rather than random, orientation to be the rule. For such studies

Figure 1.32 Bimodal orientation of carbonized wood in Pennsylvanian Shale (Zangerl and Richardson, 1963, Fig. 34). Only larger fragments are shown.

many individuals are measured, which is a time-consuming job requiring patience. Some statistics will be needed, but usually only the calculation of the vector mean, well explained by Reyment (1971, Chapt. 3) and Potter and Pettijohn (1977, pp. 374–376).

There have been numerous experimental studies of fossil orientation in flumes, such as that by LaBarbera (1977). However, most have considered sandy- rather than muddy-bottomed flumes.

When fossil debris is small, primary current orientation may be totally disrupted by bioturbation. However, we have often seen fine, invertebrate fossil debris and woody material that lack both orientation and bioturbation. Therefore, bioturbation is not always the cause of the poor orientation. Slumping should also be considered.

Larger body fossils, such as the echinoderms (crinoids, brittle stars, and starfish), with their flexible arms also have provided evidence of current direction in shales. However, the evidence is only useful to those skilled in paleoecology, who can correctly distinguish a transported dead organism from one buried in its life position. Seilacher (1973, pp. 162–165) provides a brief and excellent summary of fossil orientation and his paper (Seilacher 1960) on current directions in the Hunsrückschiefer, a Devonian black shale in Germany, is a classic. His beautiful illustrations should be consulted to see how much paleocurrent information can be gathered from some of the larger body fossils in shales, when one has the proper skills. An unusual example of the paleocurrent significance of body fossil orientation is provided by Wickwire (1936), who described oriented crinoid stems on a log embedded in the New Albany Shale in Indiana. The orientation of the stems implied a preferred orientation of currents.

What is known about the orientation of fine silt in shales and mudrocks? Because there is appreciable silt in many argillaceous sediments, over 300 claystones average 31% silt according to D.B. Shaw and Weaver (1965,

COPROLITE ORIENTATION

WAVE   **BIOGENIC**

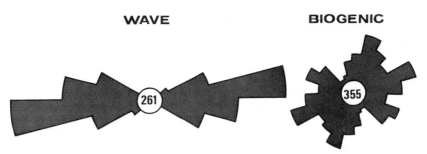

Figure 1.33 Orientation of coprolites in Rock Lake Shale (Pennsylvanian) of Kansas (Hakes, 1976, Fig. 11).

Table 4), it ought to be possible to measure quartz orientation in the plane of the bedding using low-power SEM pictures. However, we do not know of any such studies and have not yet been successful ourselves. Nonetheless, Piper (1972a, Fig. 2), using thin sections cut parallel to the plane of the bedding, found that scattered grains of silt-sized debris in laminated mudstones had a preferred orientation. Whether or not preferred orientation is present may depend on whether deposition occurred by a slowly decelerating current exerting tractive force on the interface, or by a sudden deceleration wherein all the suspended debris was suddenly dumped on the bottom.

Orientation of fecal pellets in the plane of bedding as was measured by Hakes (1976, Fig. 22) is another possibility, if either compactional distortion or organisms do not totally destroy it (Fig. 1.33). Fecal pellets are very abundant in many muds and shales. Woody debris can also be measured (Fig. 1.33).

Orientation of concretions in Devonian shales in New York has been studied by Colton (1967), who found a general correspondence between their orientation and that of sole marks. Craig and Walton (1962, Fig. 2) also reported on elongate concretions parallel to transport direction in Silurian turbidites in Wales. Orientation of elongate concretions in sandstones is much more common, however, and has been observed to parallel paleocurrent direction, possibly because the sand has maximum permeability in the direction of transport. It seems that concretions in shales are most likely to have paleocurrent significance when they form early (Fig. 1.34).

Another approach is to map the areal variation of scalar properties in a shale: percentage silt and spores or percentage of specific clay minerals such as kaolinite; or biofacies (the predominance of, say, pelagic to benthonic forams) across a shale basin; or even its organic mineral facies, insofar as the latter may be related to either provenance or to paleocirculation within the basin. All of these offer most promise only when the internal stratigraphy of the shale is well established so that correct lateral compari-

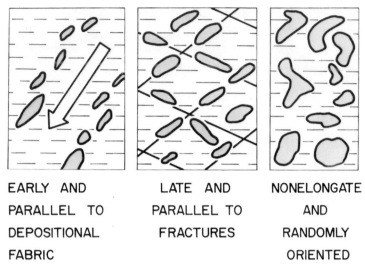

EARLY AND   LATE AND   NONELONGATE

PARALLEL TO   PARALLEL TO   AND

DEPOSITIONAL   FRACTURES   RANDOMLY

FABRIC   ORIENTED

Figure 1.34 Diagrammatic representation of orientation of concretions in shales and possible controls.

sons can be made. In the Atlantic and Pacific Oceans, Windom (1975) has shown how eolian kaolinite and quartz are dispersed oceanward by prevailing winds into deep-ocean muds deposited near the continents. In a widespread shale the sedimentologist should always try to distinguish the dispersal patterns based on eolian contributions from those based on marine circulation. The two may be largely independent, yielding different dispersal patterns.

Mapping *organic mineral associations* and relating them to provenance and/or to basin circulation is only in its infancy and few studies have been published. However, based on recent sediments, it seems that it should be possible to trace the proportion of terrestrial carbon in the organic matter of the basin. Most promising is the study of carbon isotopes (Hedges and Parker 1976, Newman and others 1973, Shultz and Calder 1976). In modern sediments the terrestrial carbon has a $\delta^{13}C$ value of about -25, while marine carbon is about -20. Some preliminary work in our laboratory suggests that in some ancient rocks this pattern may be reversed, either by diagenesis or by changes in the isotopic composition of marine carbon with time. For Devonian–Mississippian shales, we find terrestrial carbon values of about -25, but the marine carbon is around -31 (Fig. 1.35). With detailed study of this property in shaly basins we should be able to define the proportions and transport directions of the terrestrial carbon. For the best results the study of scalar properties, such as $\delta^{13}C$, should always be combined with directional structure and/or fabrics.

Study of the deep ocean and its shelves with seismic profiles, deeply towed side-looking sonar, and bottom photography has revealed mud waves from 10 to 20 m long with amplitudes of 40–50 m (Londsdale and Spiess

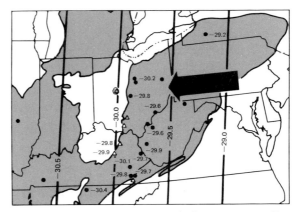

Figure 1.35 Generalized paleocurrent vector is oriented almost perpendicular to gradient of linear trend surface defined by carbon isotopes from Cleveland and Huron Members of Ohio Shale (Devonian). Linear trend surface accounts for 64 percent of total variation of δC^{13}.

1977, pp. 68–69). Mud waves with spacings of as much as 20 km and heights of 3–5 m have also been reported on some muddy coasts (Allersma 1971). All of these can be thought of as types of "megasedimentary structures," the orientations of which on the sea bottom are mappable using geophysicas and the equivalents of which should at least be looked for in ancient shales.

By combining evidence from fossil orientation, directional structures, fabric of silt, and the regional variation of scalar properties with facies distribution and paleogeography and paleolatitude, it may be possible to propose a paleocurrent system that covers much of a continent as has been done for the Cretaceous epeiric sea that covered much of North America (Fig. 1.36). We anticipate comparable maps to become much more common as sedimentologists give shaly basins the attention they deserve.

SHALE IN ANCIENT BASINS

One of the most fascinating questions pertaining to shale is how does it occur in ancient basins. This question is of particular interest because only when shale is viewed on the level of the basin, or even on a continent-wide scale, can we discern and understand its broader tectonic and environmental controls in contrast to the small-scale processes—tiny kinetic factors, one might say—discussed in the "Suspension, Transport, Deposition, and Erosion" section of this chapter.

Another way to examine the occurrence of shales is to relate them to major petrographic types of sandstones, as has been done by Weaver (1978), or even to major types of carbonate rocks and/or evaporites. Cer-

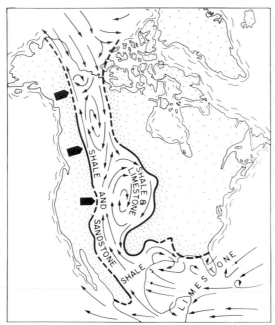

Figure 1.36 Possible generalized paleocurrent system of shallow Cretaceous seaway—a very muddy one—that extended across North America. (Redrawn from Cloud, 1971, Fig. 70.10.)

tainly this is a worthy approach, and one we hope will be further developed. However, we prefer to begin by listing all the generalizations that can be made about the occurrence and distribution of shale in ancient basins and then trying to find some unifying principles—geologic models is the current jargon—that best explain the generalizations.

Generalizations about Ancient Shales

We have gathered generalizations from both the literature and our own experience and roughly ranked them in order of decreasing importance.

- Most shales are marine.
- Thick sections of shale are much more common on the continents and their margins in geosynclines, than on cratons or in piedmont environments. This observation was made many years ago by Pettijohn (1949, Tables 131, 132, and 133). Shale is also the chief lithology of ocean basins, at least in the last few million years.
- Most thick shales in basins on the continents and their margins are related to distal turbidites and thus were deposited in relatively deep water.

- Exceptionally, shales also occur
  1. As the principal fill of long, deep channels marginal to basins
  2. As widespread, thin sheets that cover vast areas on cratons.
- Most thick and possibly even many thin shales have an internal stratigraphy defined by organic content, density, bedding, lamination, fossils, bioturbation, and mineralogy. Organic content stands foremost among these and is best revealed by color and radioactivity.
- Most thick shales and perhaps even some thin ones have a clinoform structure that is related to basin geometry and to marginal deltas and rivers.
- In the thicker shales of geosynclines, overpressuring leads to significant faulting, flowage, and mud diapirs so that the accumulation of thick shales generates its own "shale tectonics."
- Shale implies, as a rule, a wet climate and high to very high sediment supply: somewhere on the proximal landmass the source area had high rainfall and/or high relief.
- Because mud can remain in suspension for months in turbulent water, its final depositional site may have been far removed from its initial point of input along the border of a marine basin.
- The shales and subfacies of shales in a basin commonly have greater lateral continuity than their associated facies.
- Thin layers in shales can commonly be traced over very wide areas.

Discussion and Examples

Shale is chiefly related to deltaic and turbidite deposition on and near the continents and to suspension deposition in the deep ocean (Table 1.10). Hence, it is virtually always marine. Moreover, a large sediment supply, implying a high-relief landmass with high rainfall, is a universal requirement for the maximum production of mud. Such a source can be either proximal to the depositional basin or thousands of kilometers away on the other side of a continent. The initial mud fill of a shaly basin was transported to sea level either by big rivers from headwaters with high relief and high rainfall,

Table 1.10. Where and How Shale Occurs in Ancient Basins

Shelf-to-basin transitions
  Deltas prograde seaward onto oceanic crust
    Section thickens seaward to maximum of as much as 10 000 m before thinning onto abyssal plain. Full range of depositional environments from alluvial, deltaic, coastal, slope, to deep basin (abyssal). Rivers may be small or large and rate of deposition may be high or low; when high, overpressured shales and related deformation (diapirs, faulting, and folding) are common as are growth faults. Most shales are associated with prodelta muds and medial and distal turbidites. Areal extent of clastic wedge depends on abruptness of tranisition from continental to oceanic crust and water depth. Radial dispersal patterns are common in fans at base of slope.
  Deltas prograde onto craton
    Thick section along or proximal to cratonic margin with complete range of environments from fluvial and piedmont to coastal, shelf, slope, and deep basin. Shale section is best developed on delta front and slope environments, where thicknesses in excess of 2000–5000 m are not uncommon as is turbidite deposition. Moderate or little overpressuring and penecontemporaneous deformation in shale. Cratonward, shales thin but may extend over a large part of a craton as thin marine units. Submarine fans rare; paleocurrents very uniform.
  Cratonic deltas
    Delta migration, 500–1000 km or more up- and downdip, produces a thin but widespread deposit within which very thin marine shales, one to a few meters thick, are also very widespread. Full range of depositional environments from alluvial, deltaic, coastal, shelf, and broad, shallow basin. Deltas are supplied by small as well as very large rivers headwatered in highlands marginal to the craton. Neligible overpressured shale and penecontemporaneous deformation within the shale, except locally under rapidly deposited sandstones. Little or no deposition of mud from distal turbidites. Paleoccurrents may be very complex, and reflect both down dip and longshore transport parallel to depositional strike.
  Carbonate rims
    Abrupt termination of a carbonate platform. Associated shale, with very abrupt facies change, occurs laterally in a trough or basin. Paleorelief along rim may be hundreds of meters. Mud was supplied to the trough mostly longitudinally. Typical occurrence is along a quiescent continental margin or along land masses having a tropical climate, where carbonate production is high. Common in Phanerozoic time. Shale may be 1000 m or more thick.

Table 1.10. *Continued*

Rifts

    Range in size from a few hundreds of kilometers to almost continental or oceanic dimensions and may be filled with both marine and/or non-marine clastics supplied by either a large river flowing "downrift" or by small marginal ones. Shale fill is maximal, when rift is largely below sea level, climate is wet, and supply is by small muddy, marginal rivers. Shales may be 1000 m or more thick.

Island arcs

    Small to moderate sized, tectonically very active basins mostly with local supply from volcanic terrains. Mud-cored anticlines and diapirs common, as well as growth faults. Supply by small coastal rivers. Turbidity currents major factor in deposition. Mud flows and olistostromes. Shales may be thousands of meters thick.

Deep Oceans

    Thin, but very far-ranging units, mostly deposited at very slow sedimentation rates from long-traveled suspension, some of which may be from air. May be interbedded with volcanics near medial ridges or marginal trenches. Below compensation depth, shale is notably carbonate free, may be siliceous, and is highly oxidized and enriched in manganese. Appears to be rarely preserved in the ancient.

such as the bordering well-watered mountains of a large continent, or by many small rivers that drained a well-watered island arc or a mountainous, well-watered continental margin whose continental divide was nearby. One point that seems to emerge is that shale is largely a "wet climate" sediment.

    We suggest that the four major settings of ancient shales, shelf-to-basin transitions, rifts, island arcs, and deep oceans, account for 80%–90% of all the occurences of ancient shales (Fig. 1.38). Let us consider each of these in turn.

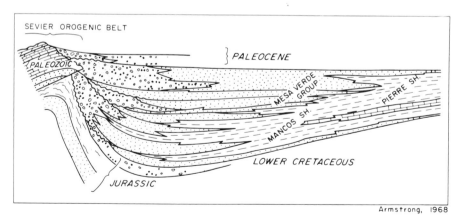

Armstrong, 1968

Figure 1.37 Downdip cross section of Cretaceous transition from conglomerate → sandstone → shale in western United States. (Redrawn from Armstrong, 1968, Pl. 35.) Probably a majority of widespread shales on cratons have had similar origins—and thus are distal molasse deposits.

Shelf-to-basin transitions include both deltas and carbonate rims. Large deltas or delta complexes may prograde cratonward from a marginal source, perhaps fed from mountains resulting from continental collision, or they may prograde seaward along a continental margin from continental onto oceanic crust. In either case, turbidites and shale will form part of a thick delta front and/or basin-plain fill in front of the subaerial part of the delta. Such shaly basins may be a few hundred to thousands of meters thick. Examples of large deltas that have prograded cratonward include the Upper Devonian of the Appalachian Basin (Oliver and others 1967), the Upper and Middle Devonian of the Franklin Geosyncline in the Canadian Arctic Island (Embry and Klovan 1976), and the Cretaceous east of the Sevier orogenic belt where the Mancos and Pierre Shales lie seaward of the Mesa Verde Group (Fig. 1.37), to name but a few in North America alone. Examples of delta complexes that have prograded seaward over a hinge line that may in part separate continental from oceanic crust include the Tertiary Wilcox of much of Texas and Louisiana (W.L. Fisher and McGowen 1967, Fig. 3), the Pliocene and Pleistocene clastics of the ancestral Mississippi River (Woodbury and others 1973, Figs. 7 and 8), the Niger Delta (J.R.L. Allen 1964, Weber 1971, Fig. 26), and many of the basins that belong to the "Tertiary delta" classification of Klemme (1971). One of the most spectacular examples of this type is the delta of the Ganges–Brahmaputra, in which most of the detritus comes from the Himalayas. These two rivers supply vast volumes of mud, silt, and sand (their headwaters in the Himalayas have both very high relief and high rainfall) to a submarine fan that is almost 2000 km long (Curray and Moore 1974, Fig. 14). Most such large river systems appear to be localized by deep-seated faulting oriented at a high angle to continental margins. Deltas from much smaller rivers can also supply mud and, where the continental shelf is narrow and not scoured by strong currents, this mud accumulates as shaly clinoform deposits on the continental slope. Turbidity currents are commonly a major factor in the transport of such muds. Much of the Mesozoic and Tertiary continental margin of Brazil, south of the mouth of the Amazon River, has shale dominantly of this origin (Asmus 1975). Likewise, consider another example: the Bokkeveld Group of Devonian age in South Africa thickens seaward to over 12 000 ft, where it also has its greatest proportion of shale (Fig. 1.39), and is now along the present coastline of a trailing continental margin. Does this indicate that the Devonian continental margin in South Africa was not too dissimilar to that of today?

Deltaic deposits, such as the Carboniferous coal measures, also occur as thin deposits that may be widespread or localized and may be limited to nearby marginal mountains or extend thousands of kilometers across wide cratons. Such deposits may have had short and small or very, very long river systems. In either case, shale is almost always a major lithology of such deltas (Fig. 1.40). In a broad sense, most of these deltaic deposits are molasse and are the result of low-energy deltaic deposition; they therefore have

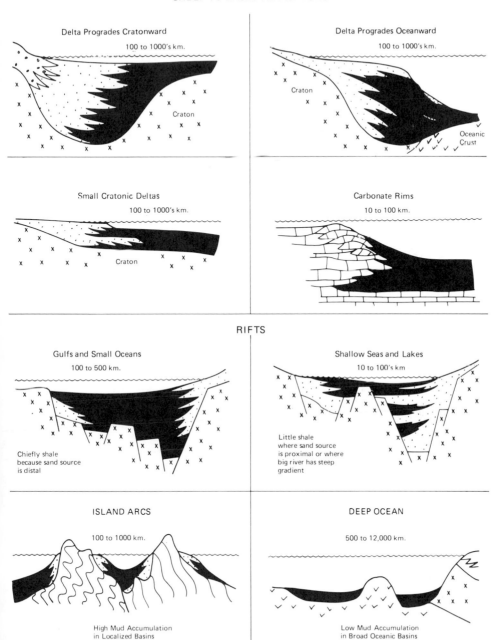

Figure 1.38 Diagrammatic representation of occurrence of shale in ancient basins.

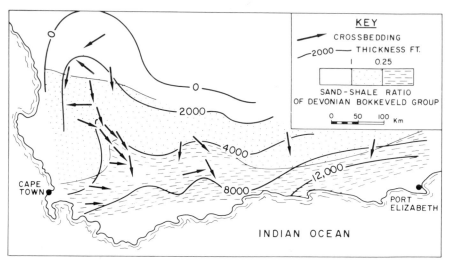

Figure 1.39 Paleocurrents and sand-shale ratio in the Devonian Bokkeveld Group of the Republic of South Africa. (Redrawn from Theron, 1970, Figs. 4 and 5.)

much associated shale, as much as 40%–70%, because the wave energy of inland epeiric seas was normally small. However, evidence for turbidity current deposition of the delta front muds in such deposits is usually hard to find (although it is also rarely looked for). The Carboniferous of North America contains many good examples of such occurrences of shale (Ferm 1970, Wanless and others 1970, Donaldson 1974, Pryor and Sable 1974). Examples are usually easy to find in the successor basins of most mountains chains, the fill of which was deposited at, near, or below sea level (Fig. 1.41). Shale can also occur in successor basins formed well above sea level but only when there is some type of "tectonic dam" that creates an artificially high base level, a situation that may be fairly common in intermontaine basins far from the world ocean.

At this point we should stop and note one of the problems associated with

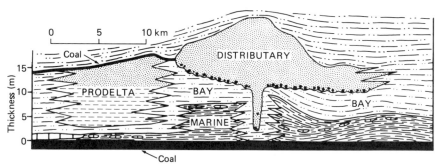

Figure 1.40 Lateral relationships of marine, prodelta and bay shales above the Hinman Coal and Lost Creek Limestone, Breathitt Formation, Pennsylvanian of eastern Kentucky.

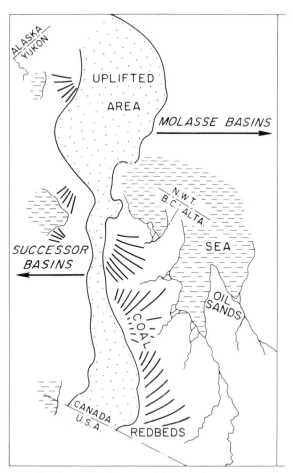

Figure 1.41 Shaly molasse, mostly deposited in prodelta environments of Cretaceous age in Canada. (Redrawn from Eisbacher and others, 1974, Fig. 8.)

trying to link shale occurrence with deltas, or with any other major environmental or tectonic feature, for that matter. This is the small fall velocity of mud particles (Table 1.3), which permits the smallest turbulence to keep them in suspension and to carry them great distances from their initial input along the margins of a marine basin. This problem is most readily perceived in deltaic deposition, especially when we stop and remember the distinction between wave dominated high energy deltas and fluvial-tidal low-energy deltas (Wright and Coleman 1973, Fig. 18). The former have much sand and protrude but little seaward from the coastline, whereas the latter have much more mud in the subaerial part of the delta and prograde seaward with long delta fingers. Given equal riverine turbidity and discharge, the rivers of high- and low-energy deltas will both carry the same volume of mud *but it will be deposited in markedly different places and environments.* In low-

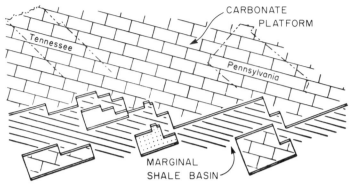

Figure 1.42 Diagrammatic representation of a wide carbonate platform with abrupt transition to shale along eastern North America in the Cambro-Ordovician. (Redrawn from Thomas, 1977, Fig. 11B).

energy deltas much mud is deposited as floccules in the prodelta and subaerial parts of the delta or nearby. In high-energy deltas most of the mud is distally deposited and, in fact, may be deposited hundreds of kilometers away, either along a coastline or in deep water on a submarine fan or cone at the base of a continental rise. Much of the mud from the Amazon, for example, is carried as much as 1600 km northwestward along the coast line of the Guyana Shield (Allersma 1971, Gibbs 1977). Likewise, mud may be washed over the shelf margin and then carried along a deep trough, one side of which might be bordered by steep carbonate reefs. Study of paleocurrent systems plus a knowledge of regional paleogeography is always vital to better understanding the occurrence of shale in ancient basins.

Shales associated with carbonate rims are less well known and are probably of lesser volumetric importance than those associated with deltas. Thomas (1977) has shown the wide extent of a carbonate–rim-shale transition in the Cambro-Ordovician along much of southeastern North America (Fig. 1.42). The famous Cambrian Burgess Shale of British Columbia, a basinal turbidite sequence (Piper 1972b), is part of the same carbonate–shale transition and together these two studies show the vast areas, virtually entire continents, over which such transitions can extend during periods of little tectonic activity, provided, of course, that the climate was warm. Another example is the Tertiary of Acquitaine, where platform margin carbonates pass abruptly seaward and slopeward into shale (Fig. 1.43). Galloway and Brown (1972, Fig. 21) show a good example in the Pennsylvanian of north-central Texas. Still another is provided by Rose (1977, p. 164) for the Mississippian of Montana and Idaho. Sandstone may replace shale as the fill of the shelf-margin trough and where this occurs there probably were channels that transported sand seaward across the shelf. Such a process seems most implausible, however, for shale-filled troughs bordering a carbonate shelf. Would not turbid, muddy waters on a shelf effectively inhibit carbonate deposition?

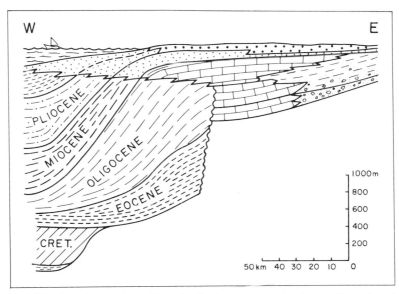

Figure 1.43 Shelf to basin transition in Cretaceous and later sediments of Aquitaine Basin of southwestern France. (Redrawn from Kieken, 1973, Fig. 3.) Similar clinoform relations probably prevail for the majority of shaly basins.

Rifts can also contain significant quantities of shale provided there is either a large, muddy, low-gradient river debouching at one end or many small rivers that supply sand only to the margins of the rift. Still another possibility is for the rift to be sufficiently wide, say 100 or more kilometers, that only mud carried in suspension can reach its central portions. Conversely, a rift lying between two low-lying arid landmasses is unlikely to be filled largely by sands and gravels. Mud accumulating in rifts may be either lacustrine or marine, the former being most likely in the early stage of rifting prior to more distal separation of continental segments. During rifting, growth faults and horsts within the rift can be active and produce penecontemporaneous deformation of the contained shale section including mud diapirs, especially where mud accumulation was very rapid. Along bordering growth faults, there may be deep-water conglomerates encased in shales as well as exotic blocks or wild flysch. Although there is still much to learn, it appears that widespread, dark, organic-rich shales were deposited in the latter half of the Mesozoic as the North Atlantic opened (Ryan and Cita 1977, Fig. 1). Surely this is the archetype of a very large shale-filled rift! Precambrian occurrences of shale-filled aulocogens have also been noted (Fig. 1.44) and, in the rift-related Triassic of the eastern United States, shale is the principal lithology (Van Houten 1969, p. 314). Sengör and others (1978) and Burke (1977) provide recent reviews of rifts in which they distinguish between those formed at random along a continental margin and

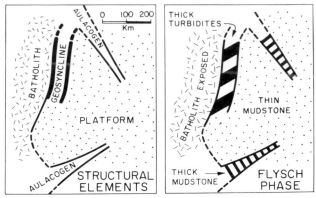

Figure 1.44 Thick Precambrian mudstones accumulated in an aulacogen in the Canadian Shield—a good example of how large scale rifting can provide a quiet, protected basin for mud accumulation (Hoffman, 1973, Fig. 24).

those formed by colliding continents. Burke recognized 13 types of rifts. How many of these 13 types have significantly different kinds and volumes of shales? Schneider (1972, p. 110) suggested four stages of the filling of a rifting ocean basin. A careful comparison of the kinds of shales in these different stages seems well worth doing, especially if the climate of the different stages is taken into account.

Comparatively little is published about shale in island arcs, in part because not too many of them have been fully explored and well understood. Present knowledge is very incomplete. Using Indonesia as the archetype, however, it appears that small to medium shaly basins may exceed thousands of meters in thickness, probably contain many mud-cored anticlines and diapirs, and have nappes that are, in part, "shale lubricated." In such island arcs penecontemporaneous deformational structures, such as growth faults, diapirs, mud-cored anticlines, and possibly even olistostromes, are all favored by overpressured shales that result from rapid deposition. Overpressured shales are an especially "mobile paste" for basins that are experiencing deep-seated tectonic deformation. Mud production in an arc will be maximal when it lies in the tropics and minimal when it lies in an arid zond. In either case, small rivers supply most of the mud to individual basins within the arch, although wind-blown volcanic ash can be a major component (Eaton 1978, Fig. 3).

The occurrence of shale in ocean basins is still poorly known and consequently our discussion is limited to the summary in Table 1.10 which is chiefly inferred.

Finally, we should mention a few famous shales (Table 1.11). Examination of Table 1.11 shows these shales to be distinctive because they are important stratigraphic units, widespread or thick. In addition, many are famous for their fossils, unusual mineralization or high organic content.

Table 1.11. Famous Shales

Alum
  Late Cambrian, organic-rich, oil shale that has also supplied some uranium. Occurs over much of Scandivavia. Olenids are its chief fossils.
Burgess
  A thick, dark gray to black, uniquely fossiliferous (well-preserved soft-bodied animals), Middle Cambrian shale marginal to a carbonate platform.
Cody–Pierre–Lewis
  Cretaceous of western interior United States extending from Colorado into the prairie provinces of Canada.
Figtree
  Precambrian of South Africa. Well known for studies of its early fossils.
Gothlandian
  Silurian of western Europe, much of northern Africa, and the Persian Gulf (northern Sweden to Mauritania, Spain to Saudia Arabia). Dark gray shales with a pelecypod and graptolite faunal association.
Green River Oil Shale
  Eocene of Colorado, Utah, and Wyoming. A highly dolomitic lake deposit that is famous as an oil shale and for its fossils.
Hunsrückschiefer
  Devonian of the Rheinisches Schiefergebirge, an intensively studied black shale with well-preserved fossils.
Kupferschiefer
  Permian of western Europe, very thin but widespread and famous for its syngenetic mineralization of Cu as well as Pb and Zn.
Mancos
  Cretaceous of western North America (New Mexico to Saskatchewan and Alberta). Thins eastward onto craton from a thick geosynclinal section.
Maquoketa
  Thick and extensive Ordovician shale that is present over much of the midwestern United States and is in part correlative with the Martinsburg. Famous for its pyritized fossils.
Martinsburg
  Well-studied Ordovician formation that passes from turbidites in the central and eastern Appalachian Basin westward into carbonates on the craton.
Mowry
  Cretaceous of western interior of the United States; widespread, thick, and siliceous. Major oil source bed.
Nonesuch
  Lake Superior region; the Nonesuch is known for its copper mineralization. Precambrian.
Ohio–Chattanooga–New Albany–Exshaw–Percha–etc.
  Devonian–Mississippian black to dark shale that covers much of North America and has much gas production in the Appalachian Basin.
Ponta Grossa
  Middle to Late Devonian of Paraná Basin in southwestern Brazil and nearby Paraguay, Uruguay, and Argentina. Thickens in Bolivia to over 5000 m.
Porters Creek
  Paleocene of the United States Gulf Coast and notable for its thickness, great extent, and abundance of montmorillonite.
Stanley
  Mississippian of the Ouachita Trough in southwestern United States. Thick flysch sequence well known as a turbidite basin.

Scott W. Starratt
Dept. of Paleontology
U. C. Berkeley
Berkeley, Ca. 94720

CHAPTER 2
QUESTION SET

Good questions are the essential starting point of all successful research.

INTRODUCTION

The methodology of shales—what to observe and measure and what does it mean—is, as we have seen, still very fragmentary and incomplete. Below is a series of questions that we have developed with the hope of advancing our understanding of shales (Table 2.1). These questions are designed to provide an inventory of essential information on which an interpretation can be based. Unlike the question set of a zoologic key, which is directed to identification, our question set is designed for *understanding.* Our list of questions is modeled after one for sands and sandstones (Sedimentation Seminar 1978), although we were initially stimulated by a series of questions asked by Wilson (1975) in his excellent book, *Carbonate Facies in Geologic History* (pp. 60–75). Earlier Folk (1968, pp. 133–138) and Ager (1963, pp. 317–318) also developed question sets. Ager, Folk, and Wilson are, to our knowledge, the first to explicitly ask questions to develop an insight into sediments; and, from a slightly different viewpoint, Spencer's (1974, pp. 788–802) structural and tectonic inventory of the world's Mesozoic and Cenozic fold belts is also very close to the theme of our question set. Our teaching experience shows that asking questions is very beneficial largely because good questions are intellectually stimulating to students young and old.

   As with the study of every kind of rock, one needs to look at it, for best results, from many different viewpoints, from the micro through meso to macro scales. Hence, below we have divided our questions into three groups: the description of cores, outcrops, cuttings, and use of wire-line logs; laboratory studies; and those larger features that can only be mapped and studied across the basin. Of course, many of the questions also relate to physical properties, geochemistry, and paleontology—all three of which play a major role in the study of any sediment.

Table 2.1. Fundamental and Recurring Questions

Describing outcrops and cores and using wire-line logs
  1. Where is the section?
  2. What are the major units?
  3. What is to be done next?
  4. What should be described?
  5. What terms should be used in the field for the description of the major lithologic types of shales?
  6. How should the observations be recorded?
  7. What paleontologic observations can and should be made in the field?
     a. Relative abundances of different macrofossils?
     b. Is the distribution of macrofossils patchy, uniform, or random? Are they concentrated within beds or on bedding planes?
     c. Are the macrofossils intact and well preserved or fragments and worn? Molds or casts? Recrystallized or replaced?
     d. What functional types of organisms are present? Encrusters, sediment trappers or binders, epifauna, or infauna? Mobile or sessile? Suspension feeders, deposit feeders, scavengers, or predators?
  8. What is the gamma-ray profile—what is it good for and how is it obtained?
  9. Wire-line logs—what are they and how can they be best used?
  10. Why and how to describe cuttings?
Laboratory studies
  1. What samples should be selected?
  2. What tools and techniques should be used?
  3. What sequence of study should be followed?
  4. How should the observations be recorded?
  5. What components are present, what is their abundance, and what do they all mean (Fundamental to the understanding of every rock, the key questions are always the same)?
     a. Large detrital grains, such as quartz, feldspar, micas, heavy minerals, and carbonate grains?
     b. Detrital clays?
     c. Authigenic grains, including carbonates (calcite, dolomite, and siderite), quartz, feldspar, zeolites, and the authigenic clays, glauconite, sepiolite, etc.?
        i. How are authigenic minerals recognized in a mudstone or shale? Those formed after deposition either by precipitation in pores or by transformation of original detrital minerals (by solution and replacement).
        ii. Significance?
     d. Mineral cements, such as silica, carbonate, or zeolites?
     e. Floccules?
     f. Pellets and pelaggregates?
     g. Organic particles?
     h. What can be learned from micropaleontology and palynology?

Table 2.1. *Continued*

6. What textural parameters should be measured and what do they mean?
   a. What proportions of clay, silt, and sand?
   b. Percentage of large micas?
   c. Size ranges and modes of diverse detrital grains?
   d. How well oriented are the different components of the shale sediment—the clay minerals, the silt and sand grains commonly found in them—and what significance, if any, does their orientation have for paleocurrents?
      i. Perfection of framework orientation?
      ii. Silt and sand grains?
   e. Burrowed and/or mottled textures?
   f. Pore geometry—kinds and amounts?
   g. Is there any significance to the shape of shale cuttings?
   h. What can be seen by radiography?
7. What name is to be used and how should the shale be classified now that we know so much about it?
8. The petrographic report: What is the best way to organize the foregoing petrology and texture into a useful, concise, and coherent petrographic report?
9. What can be learned from the study of inorganic geochemistry?
   a. The major elements?
   b. The trace elements?
   c. Exchangeable cations?
   d. Pore-water chemistry?
   e. Stable isotope geochemistry?
10. Organic chemistry?
    a. What are the best indicators of thermal history?
    b. What does the study of paleobiochemical indicators tell us?

Making a synthesis and basin analysis
1. Where did the mud come from?
2. How was it transported to its final depositional site?
3. At what water depth, sedimentation rate, oxygenation, and toxicity to life was the mud deposited?
4. What has happened to the mud since deposition?

DESCRIBING OUTCROPS AND CORES AND USING WIRE-LINE LOGS

From almost any point of view—paleontology, geochemistry, civil engineering, or stratigraphy—the field description of cores and outcrops of mudstones and shales is absolutely fundamental, for, if it is done incorrectly, very little else can be expected to turn out well. Lithologic description of cores and outcrops defines the internal stratigraphy of a shale, provides a first (perhaps the best?) appraisal of its depositional environment, and is the basis for interpreting wire-line logs and even seismic cross sections. In his paper

"Why Study Cores?" Harms (1970) effectively summarized the reasons for studying cores, one of which is their value for prediction. His arguments apply equally well to the careful description of outcrops. As in most aspects of successful field geology, the best success is to be had by consistent, perceptive observation using only the most simple of techniques—your eyes, your teeth (to test for silt and sand), a hand lens, a bottle of index oil, a Brunton, an acid bottle, a color chart, and a meter stick. To this simple list we can think of only one sophisticated addition—a scintillometer or geiger counter to obtain a continuous gamma-ray profile of the core or outcrop. Of all of these, we rate the eye—with its link to the brain—by far the most important, for it is this pair that chiefly generates our perception of reality and how to interpret it. It is also this pair that distinguishes between *seeing* and *observing,* between simply *looking at things* versus a *conscious awareness of their specific properties.* To put it differently, the beginning student *sees as much* as the old seasoned hand, but it is only the latter who *observes carefully* and thus can begin the *creative process of thinking and interpretation.*

You may find it useful to take with you into the field one of the several books or manuals on field geology (Berkman and Ryall 1976, Compton 1962). We have always benefited by this.

1. Where Is the Section? If a precise knowledge of the location of your section is lacking, it is rarely worth careful study. Hence, when you have selected an outcrop, the first step is to plot it carefully on a field map and describe its location, either with respect to a grid or to some nearby geographic landmarks—preferably both, because the landmarks can guide a stranger to it much more easily than grid coordinates alone. If you are studying a core, be sure you understand its location and have a complete description of its name and elevation.

Grid coordinates are most helpful for latter transfer to different scale maps, but your description of the section should also be related to landmarks, such as buildings, trees, gullies, prominent resistant benches, or possibly roadside distance markers. If this is done, a stranger can recognize at a glance what you have described. Also, of course, good photographs are very useful for more precise location.

2. What Are the Major Units? Make a reconnaissance of the section (either the outcrop or the entire length of core) to establish its major units, mostly on the basis of color, texture and kinds and abundance of interbedded lithologies. In outcrops, the weathering expression is very useful in this regard and should never be ignored. In other words, weathering detects and amplifies very subtle lithologic differences that might well be passed by when examining a core. By doing this, one establishes, at the very start, *the largest, most significant, and most obvious units to which all the myriad other details of the section can be related.* It is these initial big units that become the major "bundles" of your description. Additionally, by defining these major subdivisions, you can best plan your sample collection for later laboratory study and how best to allocate your time. By making such a reconnaissance you also discover and avoid any complications resulting from faulting or folding. When you are describing a core, it is essential that you understand the numbering system on the boxes and the core sequence within the boxes. This is especially necessary with shales, because they have such a uniform appearance. Commonly, arrows are used to indicate top or bottom.

3. What To Do Next? Measure and mark off the outcrop or core in 1- or 2-m units or larger, depending upon the outcrop's or core's length and your time. Each major subdivision is then described in whatever detail you desire.

4. What Should Be Described? Shales, as do all other natural objects, have a virtually limitless number of attributes to be described, depending upon your training, experience, objectives, and imagination. To this we should add that it is very common to describe only what one is looking for or what one has already experienced.

We suggest that a check list, even a simple and short one, can be very helpful in answering the question of what to describe (Table 2.2). This check list will also be very useful in the study of cuttings under the binocular microscope.

The application of a few drops of mineral oil to the fresh surface of a shale greatly aids its examination and description. Mineral oil is widely used in field and laboratory studies of carbonates and sandstones.

5. What Terms Should Be Used for the Field Description of the Major Lithologic Types of Shales? For field use one can use the 11 terms tabulated in Table 2.3, terms that we have taken, and in part adapted, from *The Glossary of Geology* by Gray and others (1972) or follow the terms in our own Table 1.1. All of the terms of Table 2.3 may be modified by color (chiefly black, gray, dark gray, greenish gray, maroon, red, green, etc.), or by the presence of minor minerals, such as calcite, dolomite, and/or gypsum, and by structural terms, such as "well fractured," "hackly," and "slickensided." Later in the laboratory a more refined classification can be used after petrographic, engineering, or even perhaps geochemical study.

We call your attention to one problem very common in outcrop descriptions—weathering—and how it may change your description and naming of the shales. The best way to treat this problem is always, of course, to obtain as fresh a sample as possible and always to compare the fresh sample with its surface-weathered equivalent. Digging back into the surface of an outcrop will commonly show that much, if not all, of its fissility is a product of weathering.

6. How Should the Observations Be Recorded? Field notes are perhaps the most important record of descriptive activities and are the

Table 2.2. Checklist for Description of Shales

Attribute	Remarks
Rock name	You may have to examine the sample carefully using a hand lens and even taste it to assign the correct rock name (see "Classification," p. 12–17).
Color	Use a color chart; several are available (Goddard and others 1975, Kelly and Judd 1976, Munsell Color Co. 1975). Note whether the sample was wet or dry when color was described.
Induration	Can be a weathering characteristic and described as resistant (hard) or nonresistant (soft).
Fractures..........	Hit with a hammer and specify breakage pattern—concoidal, hackly, blocky, brittle, splintery, etc. Fracture characteristics are especially useful in the study of shale cuttings (Atherton and others 1960).
Grain size	Rough estimate of amount of clay, silt, or sand. Your front teeth are very sensitive to even tiny amounts of silt and sand, or test for grit and clay content between your fingers. A hand lens and grain comparator may be useful, especially if you apply some mineral oil beforehand. Mineral oil helps reveal the finest details of texture (see also Table 2.4).
Lamination and .... bedding	Observed both with the eye and with hand lens on weathered as well as fresh and broken surfaces. Both in cores and in outcrops, wetting or oiling greatly enhances your ability to discern details. Specify some typical thicknesses (see also Table 2.4).
Fossils	Types, abundance, condition, and orientation. *Never pass these by* and if you do not know them, collect them for someone who does.
Bioturbation	Kinds, amounts, and orientation, vertical or parallel to bedding. If possible, determine if structures occur within a bed, say a thin siltstone, or on its top or at its base (see also Table 2.4).
Organic Content...	Kinds and amounts.
Accessory minerals	These include calcareous and gypsiferous cements and nodules as well as pyrite and micas plus silt- and sand-sized quartz, feldspar, and glauconite.

primary basis for correlation, facies analysis, isopachous and lithofacies maps, and especially for the spatial location of all samples to be used in petrologic and geochemical analysis. They are the permanent record and reference for future use and, hence, should be clear, concise, and *legible*. Each page of the field notes must be correctly labeled, sequentially numbered, and dated. It is also a good practice to make a copy for your permanent files after each field visit, just in case the field notebook is lost, falls into a creek, or is stolen. Also, of course, more and more your description may

Table 2.3 Common Field Terms Used to Describe Shales (Adapted from Gary and others 1972)

Clay . . . . . . An unconsolidated generally plastic argillaceous sediment rich in clay minerals and silt. When thoroughly wet, forms a pasty, plastic moldable unpermeable muddy mass. Most of the particles of a clay will be smaller than 4 μm (p. 130).

Claystone. . An indurated clay having the texture and composition but lacking the fine lamination or fissility of shale; a massive mudstone with clay predominating over silt (p. 131).

Clay shale . Shale that consists chiefly of clayey material and becomes clay on weathering (p. 131).

Clay slate. . Low-grade slate that differs from an argillite because it has parting, slaty cleavage, or incipient foliation (p. 131).

Mud. . . . . . . Slimy, sticky, slippery mixture of water and silt and clay, the consistency of which varies from a semifluid to that of a soft, plastic sediment (p. 468).

Mudrock. . .Synonym used for a mud-supported carbonate rock. Also the plural is a loosely used synonym for argillaceous rocks.

Mudstone. . An indurated rock consisting chiefly of fine detrital silt, mostly quartz, followed by clay minerals, but lacks fine lamination or fissility; in other words, the blocky or massive equivalent of a shale.

Shale . . . . . . Fine-grained, indurated, finely stratified (laminae 0.1–0.4 mm thick), and/or fissile rock commonly consisting of silt and finer clay minerals, the silt predominating and being chiefly detrital quartz. Shale is generally soft enough to be scratched but sufficiently indurated so that it will not fall apart on wetting. Has smooth feel and a splintery fracture. See Tourtelot (1960) for a history of the term (p. 649–650).

Argillite . . . .Compact rock, derived from an argillaceous sediment, that is more indurated than either a mudstone or shale and lacks the fissility of shale or the cleavage of slate. Also sometimes used for a weakly metamorphosed argillaceous rock (p. 37).

Slate . . . . . . Compact rock with fissility or cleavage independent of original bedding and formed by the metamorphism of mudstones, shales, etc. (p. 664).

Phyllite. . . . An argillaceous, indurated metamorphic rock with a silky sheen on cleavage or schistosity surfaces (p. 537).

be ultimately digitized so that graphic plots may be directly made by computer (Odell 1977).

Using the check list of Table 2.2, the section or core can then be described, detailing and summarizing the major subdivisions or bundles defined and entering these into the field notebook. Special attention should be given to laminations and thin, nonshaly laminations and interbeds for a better perception of primary sedimentation (Table 2.4) and also to help es-

Table 2.4. Laminations and Thin Beds of Sand, Sandstone, and Carbonate in Muds and Shales: Understudied Environmental Discriminators (Adapted from Reineck 1974, Table 1)

Characteristic	Relevant questions
Grading	What types—content (vertical variation of mud content within the bed) versus size grading (vertical change in size of framework) and what scale (micro or macro): Are the graded units complete or truncated (either at top or bottom?
Lamination	Good, poor, or nonexistent?
Cross-lamination (includes ripples)	Scale, kind, orientation, and directional homogeneity (unidirectional, bipolar, or random)?
Sole marks	Diverse types including flutes, grooves and bounce and brush casts, all typically of small scale but most informative. Consolidated beds only.
Erosional discordances in sand	Present or absent? If present, describe.
Lateral grain-size change	In good outcrops or in closely spaced cores can a lateral change in grain size be seen? How does its direction compare with evidence from paleocurrent indicators?
Thickness	Specify mode, minimum, and maximum and possibly determine frequency distribution of different types should they exist.
Sequence in sand	Is there a preferred internal sequence of structures and/or bedding types in the sand layer? If so, what is it?
Sand content of proximal mud	Low, medium, or high? How does it vary within and along the beds?
Sequence in mud	Do the muds above sand layers consistently differ (bioturbation, texture, and fossils) from those below? Are basal and/or upper sand–mud contacts discordant or erosional?
Bioturbation	Present or absent? Kinds and amounts? Extends through sands into muds or only confined to sands? If confined to sand layers, within or on top or bottom?

Table 2.4. *Continued*

Characteristic	Relevant questions
Fossils	If present, what kinds (for example, pelagic or benthonic forams) and are they transported or *in situ?* Are fossils oriented and how does their orientation compare to that of associated directional structures? Ecologically most diagnostic in carbonates.

timate compaction (Fig. 1.27). Experienced field workers always made a scale drawing or sketch of the section to be described (Fig. 2.1). Columnar section paper is very useful for this purpose. Summary descriptive notes are then written along the side of the sketch. In addition, any photographs and samples taken for laboratory analysis may be located on the sketch and in the notes for future reference. Table 2.5 is an example of an edited description that you may wish to follow.

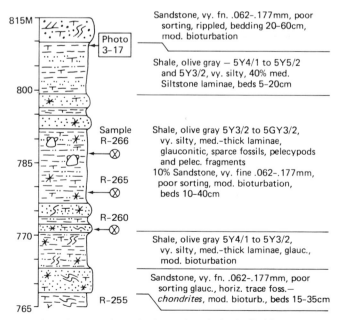

Figure 2.1 Field notes taken from a description of a shale in Pakistan.

Table 2.5 Example of an Edited Field Description (Provo and others 1977, pp. 36–37)

Berea, Kentucky, Section

Nearly complete section of New Albany Shale exposed in series of cuts on east side of Interstate Highway 75 at and south of exchange and crossing of Kentucky Highway 21 west of Berea, Madison County, Kentucky. Top of section is south of overpass, 575 ft FNL and 600 ft FWL of Sec. 7-M-63; base of section is along entrance ramp to the northbound lanes of Interstate Highway 75 (Berea quadrangle). Section measured, described, and sampled, and its radioactivity profile measured using Jacob's staff, tape, Abney level, and scintillometer by R.C. Kepferle and Paul Edwin Potter, July 9, 1976.

	Thickness (ft)

Quaternary (?)

15. Soil, olive-gray (5Y 4/1) to yellowish-gray (5Y 8/1), containing, near base, quartzite pebbles, siliceous geodes, and phosphatic nodules derived from nearby and underlying bedrock units; weathers to grassy flat. Erosional contact.   5+

Devonian:

New Albany Shale (incomplete):

14. Shale, brownish-black (5YR 2/1), weathers light gray (N7), with some iron oxide and sulfide stain on fractures and bedding planes along with rosettes and prisms of selenite 3 to 5 mm long; silt in discontinuous laminae 1 to 2 grains thick; pyrite in discoidal concretions and disseminated grains; phosphate in nodules which are round to amoebiform, 2 to 3 cm thick, elongate, some more than 13 cm across, brownish-gray (5YR 4/1) and brownish-black (5YR 2/1), weather yellowish-gray (5Y 8/1) on surface, with earthy luster and rough fracture; fossils include *Tasmanites* spores and a vitrain layer 3 mm thick 5 ft above base. Top at or near contact with grayish-green shale of basal Borden Formation seen in outcrop less than 1 mile to the west. Basal contact placed at lowest phosphate nodule.
                                                                             9.0

13. Shale, brownish-black (5YR 2/1) to grayish-black (N2), like unit above, except contains no phosphate nodules; 8 ft above base is zone of pyrite concretions 0.1 ft in diameter and 0.5 in thick, concentrated along bedding planes; shale weathers to fissile, brittle flakes and plates as much as 0.4 ft in diameter; *Tasmanites* abundant; possible fish scale, 1 mm across near top of unit; silt laminae increase in abundance downward; tough and dense, with subconchoidal fracture where fresh. Sharp basal contact.   18.0

Three Lick Bed:

12. Shale, greenish-gray (5GY 5/1–4/1), weathers yellowish-gray (5Y 8/1), clayey, subconchoidal fracture on joints; partings coated with limonite and sulfate stain. Sharp contact with underlying unit.   0.8

11. Shale, black (N1) to grayish-black (N2) and brownish-black (5YR 2/1); weathers light gray (N7), with iron oxide and sulfate stain; silt laminae 1 to 2 grains thick, commonly about 5 mm to 1 cm apart; brittle flakes litter outcrop; burrowed in upper 0.1 ft, burrows filled with greenish shale from overlying unit?; discontinuous cone-in-cone limestone layer, dark-gray (N5), 0.1 ft thick, about 0.6 ft below top (sampled); basal contact sharp.   1.2

7. What Paleontologic Observations Can and Should Be Made in the Field?

Every macrofossil and biogenic sedimentary structure seen in a shale in either a core or the field should be noted and identified or, if not identified, collected and given to someone who can do so. When collecting for later study in the laboratory, always indicate the top of the specimen and mark the north direction on it. Even if you are not well trained in paleontology, make an effort to estimate the abundance of the major types of macrofossils, even if you can only apply phylum names, such as "brachiopods," "ammonites," or "graptolites," and, of course, carefully describe their occurrence in the section.

The orientation of fossils, be it preferred or random, is also something that should be looked for and measured by every sedimentologist studying shales in either outcrops or oriented cores, for in many of the purer shales fossil orientation may be the only effective paleocurrent indicator. Orientation of graptolites (Moors 1969), possibly sponge spicules in a siliceous shale, elongate foraminifera, *Lingula,* woody debris (Fig. 2.2), or even segmented body fossils are all examples of fossils whose orientation has been used to infer paleocurrents. Seilacher and Meischner's (1964) study of fossil orientation in the Lower Paleozoic Syncline of Norway is an outstanding example (Fig. 1.5), one rich in methodology, and the study by Jones and Dennison (1970) of fossil orientation (over 13 000 measurements) in the Athens (Ordovician) and Chattanooga (Devonian–Mississippian) Shales of the Appalachian Geosyncline is an interesting application. In all such studies an effort should be made to determine whether the orientation is the result of transport after death or represents a living, *in situ* position. We stress fossil orientation, because it is something every sedimentologist can measure, regardless of his or her level of paleontologic training.

Much more fundamental to the understanding of any shale, however, is an *inventory of its macrofossils and biogenic structures and their paleoecologic interpretation.* Two general source books to paleoecology and how it can expand our understanding are Ager (1963) and Imbrie and Newell (1964). Additional references well worth your attention are Schäfer (1972) and Gall (1976). Basically, what is required of the paleoecologist is a reconstruction, insofar as it is possible, of the salinity, light, temperature, turbidity, wave and current energy, and substrate type indicated by the different macrofossils present in an argillaceous sediment. Normally, one or more skilled paleontologists or paleoecologists will be required for this task, but the sedimentologist can help significantly by his collection of specimens, by noting good collecting sites, and, perhaps most important of all, by working closely with paleontologists to fully insure that sedimentologic evidence from the shale is fully integrated with that from the fossils themselves. When combined with evidence gathered from minor interbedded lithologies plus sedimentologic features of the shale and its organic geochemistry, a paleoecologic understanding of the macrofossils can go far to complete our perception of its environment of deposition. Examples of

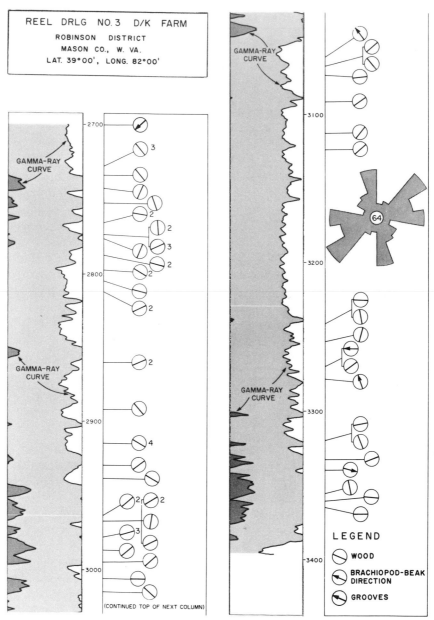

Figure 2.2 Orientation of wood, brachiopods and grooves in an oriented core from the Devonian shale sequence of the Appalachian basin.

paleontologic studies of macrofossils in shales, all of which include some paleoecologic inferences, are Hattin's (1965) study of the Cretaceous Granerous Shale of Kansas, the paleontologic reexamination by Morris (1977, 1979) of the famous Burgess Shale (Cambrian) in British Columbia,

Thayer's (1974) study of paleoecology in the Upper Devonian of New York, and the classic study of the Devonian Hunsrückshiefer in Germany by Richter (1936).

Below are some specific questions that are fundamental to an enhanced understanding of any macrofossil assemblage.

   a. *Relative abundances of different macrofossils?* Diversity generally increases seaward toward the edge of a continental shelf.

   b. *Is the distribution of macrofossils patchy, uniform, or random? Are they concentrated within beds or on bedding planes?*

   c. *Are the macrofossils intact and well preserved or fragmented and worn? Molds or casts? Recrystallized or replaced?* The condition of the fossils gives information on the energy of the depositional environment, on the *in situ* or transported nature of the fossils, and/or on later diagenetic events.

   d. *What functional types of organisms are present? Encrusters, sediment trappers or binders, epifauna, or infauna, mobile or sessile? Suspension feeders, deposit feeders, scavengers, or predators?* Animal–sediment relations influence deposition, stabilization of the water–mud interface, and mixing of the newly deposited sediment. Different modes of life and feeding habits indicate, of course, different ecologic and sedimentologic conditions.

   e. *Recurrent associations of fossils through a vertical sequence?* Their presence shows the shifting back and forth of environmental conditions.

Moreover, biogenic sedimentary structures are not to be forgotten, the basic questions being essentially all of the foregoing, (a) through (e). Good general sources for interpreting biogenic structures are Frey (1975) and Crimes and Harper (1970, 1977).

8. What is the Gamma-Ray Profile—What Is It Good for and How Is It Obtained?

A gamma-ray scan (Fig. 2.3) is an easy and quick way to see the overall stratigraphic units you are describing (Provo and others 1978). We have also found that a gamma-ray scan is an absolutely essential activity in the study of cores and outcrops, because virtually every good sedimentologic study of outcrops requires integration with subsurface geology. Moreover, the gamma-ray scan, better than anything else, facilitates direct comparison to the gamma-ray wire-line logs of the subsurface. The gamma-ray scan is best done by systemically making measurements at regular intervals. It has been our experience that the most useful profiles are based on small intervals, for instance, at regular 1-ft or 20-cm spacings along the core or outcrop, and especially at changes in lithology. Within, say, marked 1-m intervals, simply divide the interval into parts corresponding to the number of readings you desire and plot the readings sequentially, there being no need to record the elevation of each reading. When long sections are studied, it is helpful to have one person record. In sum, gamma-ray profiles are extremely useful in correlating outcrops, correlating outcrops with subsurface

Shales commonly have much more lateral continuity than sandstone or carbonate bodies. Minor breaks in sedimentation often result in widespread, thin, dewatered horizons (hardpans) that have higher densities than the underlying sediments. Such density differences may be recognized on the wire-line logs and can be used as effective correlation horizons. These, along with bentonite horizons, define approximate correlative time surfaces (Asquith 1970) within shales. Probably there are many more disconformities in shales than have been recognized and every technique must be used to recognize them (Voight 1968, Baird 1976).

A good elementary starting point for understanding wire-line logs is C.A. Moore (1963), but more complete guides are provided by the manuals of Dresser Atlas (Various Authors, 1974) and by Schlumberger Ltd; manuals of the latter can be obtained from their local company field offices, as well as from their North American headquarters at 277 Park Avenue, New York, New York 10017. A recent publication on subsurface geology is Low (1977).

10. Why and How to Describe Cuttings? Because cores, especially of shales, are uncommon—typically it is only the sandstones and carbonates or coals and evaporites that have been of interest—it may be necessary to study cuttings of selected wells to fully understand and interpret your wire-line logs.

As with outcrops, description needs to be systematic (see Table 2.2) and one needs to be continually perceptive and alert to lithologic changes. In general, work *down* the hole, recording the first appearance of new lithologies to avoid being confused by slumping from up the hole. Cuttings should be described wet—with water or index oil—in a glass or plastic tray such that all the cuttings are submerged—using a binocular microscope at low power, typically with $10-20 \times$ magnification. However, the use of oil eliminates the possibility of later chemical analyses. If they are available to you, some low-power SEM photographs taken of selected shale types will help you better identify and define the constituents of the cuttings as seen by binocular microscope. The shape of the cuttings should also be noted. For example, in the Illinois Basin chips of Pennsylvanian shales have different shapes from those of Chesterian (Mississippian) shales (Fig. 2.5). Also, remember to make use of standard field equipment, such as an acid bottle and color chart.

Three guides to the description of cuttings are Low (1951, 1977), Maher (1959), and Beckmann (1976, pp. 112–116).

LABORATORY STUDIES

Samples collected in the field and from cores or well cuttings provide the raw materials from which detailed, specific petrologic and chemical data are obtained in the laboratory. These data are fundamental to precise in-

Figure 2.5 The shape of shale cuttings can be clues to stratigraphic identity and sequence: **A.** Caseyville (Pennsylvanian), **B.** Chester (Mississippian) and **C.** selected lath-shaped Chester Fragments (Atherton and others, 1960, Fig. 7).

depth description and classification and become the basis for more refined interpretations, such as provenance, depositional processes, and diagenetic history. Because fineness of grain has historically precluded obtaining much information from shales and mudrocks in the field, laboratory study is relatively more important than for sandstones and carbonates.

As in field geology, success in laboratory studies is dependent upon precise, perceptive observations and measurements. Unlike field geology, however, more sophisticated techniques and instruments are required: microscopes (ranging from the binocular to a standard petrographic microscope to the SEM), X-ray diffraction analysis, and radiography, as well as geochemical techniques.

Petrologic and chemical data are time consuming and often expensive to obtain; hence, it is most important to devote some effort to the rational selection of samples and techniques that best fit the needs of the problem. *Decisions should be based on what you want to know about the mudrocks or shales, what equipment can be utilized, and how much time and money are available. Therefore, it is important to have a firm grasp of the problem and the questions to be asked and a definite but flexible plan to follow, so that the laboratory analysis may be completed in the most efficient and rewarding way.*

The following questions deal with treatment of samples once they arrive at the laboratory.

1. What Samples Should Be Selected? This first question is the basis upon which all following answers depend. All statistical treatment of the data will be directly influenced by methods of sample selection. Should samples be taken at regular intervals, say, every 0.5 m of section, or perhaps on some statistically random basis throughout the section? Likewise, should samples be taken at every vertical change in lithology, or should every major bundle be sampled separately at its base, middle, and top? These options are all open to the investigator and the final choice can be made only on the basis of careful planning. Some discussion of geologic sampling is given by

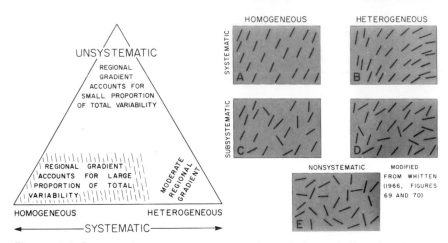

Figure 2.6 Systematic versus unsystematic variation of directional structures, say graptolites, can be expressed in terms of the strength of the
regional gradient relative to local variability. Scalar properties can be considered in the same way. (Redrawn from Whitten, 1966, Figs. 69 and 70.)

Krumbein (1960).   Whitten's (1966, pp. 70–95) discussion of the field
sampling of the structural features of folded rocks is very relevant to geologic
field sampling in general, including the sampling of shales (Fig. 2.6).   We
generally prefer a sample design that is representative of the geologic distribution of rock types to a purely random design.   Again, the sample design
must be realistic in terms of the time and effort to be invested.   Selection of
1000 thin sections at 1-ft intervals may be statistically sound, but if the job is
to be done within 1 month, it is unrealistic.

2. **What Tools and Techniques Should Be Used?** Carefully define the
questions and equally carefully select the tools and techniques that will give
the best answers within a time–cost framework.   If the answer to a question
does not require pipet analysis, then do not routinely do it.   An important
step in analyzing a suite of samples is to estimate the time required for each
analytical task and multiply it by the number of samples to determine the
projected man-hours that will be required.   This may prevent you from
overextending yourself.   Table 2.6 presents the types of techniques and tools
that may be used in petrographic analysis, the kinds of data that can be
collected, and the goals that can be achieved by their use.

3. **What Sequence of Study Should Be Followed?** As a general rule,
the samples should be split, saving a portion for a permanent reference
collection, and the splits parceled out for various analyses.   The best way to
do this is to construct a flowchart of sample analysis.   Plan carefully, to make
the best, most efficient use of the available sample material.   For instance,
destructive analyses should be scheduled last in the sequence.

4. **How Should the Observations Be Recorded?** The answer is *system-*

Table 2.6. Common Petrographic Techniques and Their Utility

Technique	Data and goals
Binocular microscope Examination of well cuttings and core and outcrop chips. Scanning electron microscope examination of selected cuttings can resolve many problems and should be done early.	General compositional description and identification, fossils → classification, paleoecology, correlation, and provenance.
Polished slabs Best coated with oil or plastic spray and stained. Usually inspected and photographed.	Bedding properties; primary, secondary, and biogenic structures; color; textures; grain size → paleohydraulics and depositional environments.
Size analysis Disaggregated, sieve, or pipet analysis, and thin-section measurements. Statistical treatment of data.	Sand–silt–clay content, size range of sand and silt particles. Generally not meaningful for clay mineral particles → paleohydraulics and depositional environments.
Petrographic analysis Thin sections and petrographic microscope. Sections generally cut perpendicular to lamination, but fossils may be seen much better if sections are cut parallel to lamination. Ultrathin sections, 10–20 μm thick are best. Colored microphotographs very useful.	Mineralogy, texture, microfossils, cements, fabric, pore space, and microlamination → classification, provenance, diagenesis, paleoecology, correlation, and depositional environments.
X-Ray diffraction Samples disaggregated, then sedimented on slides or packed in holders, or smears prepared; or thin-section blanks can be used. Sedimentation techniques can select size ranges or bulk samples may be analyzed. An inexpensive, routine analysis.	Clay mineralogy, bulk mineralogy, crystallinity, and fabric → provenance, depositional processes and environments, diagenetic history, classification, and correlation.
Radiography Slabs cut 0.25 in. or less, can use polished slabs. Radiograph photographs produced.	Primary, secondary, and biogenic structures, fossils → paleohydraulics, paleoecology, depositional environnments.
Scanning electron microscope Samples prepared, gold plated, and mounted on stubs. Visual observation and photographs. Especially good for very high magnifications.	Ideally fabrics, textures, pore space geometry, microfossils, mineralogy → depositional and compactional processes, paleoecology, and diagenesis.

atically; therefore, a report form is essential and most probably it will be necessary to make one for the problem at hand. When so doing, keep in mind that you may want a form which will readily lend itself to computerization of your laboratory data just as with your field data.

5. What Components Are Present, What Is Their Abundance, and What Do They Mean? These key questions, are fundamental to the understanding of every rock.

  a. *Large detrital grains such as quartz, feldspar, micas, heavy minerals, and carbonate grains?* These grains give the provenance of the coarse fraction of the shale just as with sandstones. Moreover, by systematically mapping composition across a basin, dispersal patterns may also be recognized.

  b. *Detrital clays?* The three main groups of detrital clays are kaolinte and halloysite (1:1), the smectites, illite, etc. (2:1), and the chlorites and vermiculites (2:1:1). The mineralogic identification of the clay minerals is the most standardized and widely used technique in the study of shales; four useful references are Biscaye (1965), Carroll (1970), Cook and others (1975), and Grim (1968). Mapping clay mineral ratios may help outline the shoreline of a marine basin (Fig. 2.7), possibly provide a better understanding of how different injected fluids may react with a shale, and also explain many physical properties. The x-ray identification of clays in thin section should be very helpful (Wilson and Clark 1978).

  c. *Authigenic grains including carbonates (calcite, dolomite and siderite), quartz, feldspar, zeolites, and the authigenic clays glauconite, sepiolite, etc.?* There are two distinct aspects of this question—recognition and significance.

    i. *How are authigenic minerals recognized in shale—those that formed after deposition either by precipitation in pores or by transformation of original detrital minerals (by solution and replacement)?* This is one of the most difficult of problems in sedimentology. Authigenic minerals may occur as replacements (pseudomorphs) of original detrital grains, as new materials (crystals or concretions), and as cement. Some aids to the identification of authigenic minerals in shales include exotic mineralogy (e.g., zeolites), euhedral form, and grain size much larger than that of the shale's framework. Nonetheless, one of the major problems of shales resides right here—distinguishing an authigenic from a detrital mineral or, in other words, being able to draw a paragenesis diagram for a shale.

    ii. *Significance?* Those minerals that form shortly after deposition may be good estimators of the chemistry of the original water column in the sedimentary basin. For example, the series pyrite–siderite–glauconite probably indicates increasing oxygenation and pH (Garrels

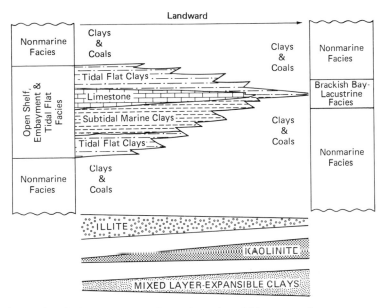

Figure 2.7 Lateral relationships of clay minerals within a marine-to-non-marine pinchout, Desmoinesian Series (Pennsylvanian), northern Missouri and southern Iowa (Brown and others, 1977, Fig. 14).

and Christ 1965, p. 228). Sepiolites and zeolites, in contrast, indicate high silica in pore water (Stonecipher 1976). Still other authigenic minerals, such as the feldspars and sulfates, are difficult to interpret.

d. *Mineral cements, such as silica, carbonate, or zeolites?* In many mudstones and shales it is virtually impossible to distinguish in thin section a small detrital grain from a mineral cement, unlike thin sections of sandstones. Perhaps SEM study with an X-ray analyzer may significantly advance this important problem.

e. *Floccules?* Floccules are often destroyed during early compaction but, when preserved, are best seen in thin section and SEM. They are recognized as small domains of clay- and silt-sized particles (Pryor and Van Wie 1971). The size, distribution, and mineralogic makeup of floccules may be indicators of depositional conditions. More research is needed to observe and interpret floccules.

f. *Pellets and pelaggregates?* Indicators of significant organic productivity, these are best seen in thin section as small textural domains defined by fabric, organic content, and color (Fig. 2.8). In some shales they are totally absent but in many, especially in marine muds and shales, pellets and pelaggregates may be the dominant components (Pryor

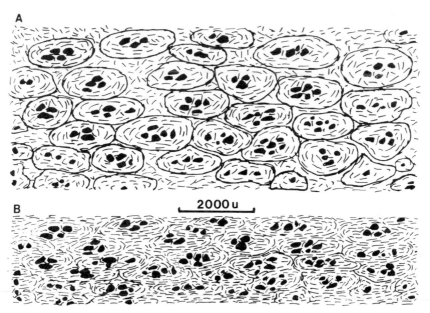

Figure 2.8 Pellets and pelaggregates: **A** at time of deposition and **B**, as seen in thin section.

1975). Most sedimentary glauconites derive from the diagenesis of pellets and pelaggregates and indicate relatively slow sedimentation. In muds, pellets and pelaggregates differ from floccules in their higher organic content and their larger size (pellets and pelaggregates range from 0.25 mm to more than 2 mm, whereas floccules average about 100 μm). In shales, pellets and pelaggregates are much harder to see than in carbonates, because in the latter cementation is early and, consequently, compaction is minimal and many more original structures are preserved.

g. *Organic particles?* Under this heading we include spores and woody fragments as well as amorphous material. The ratio of woody to other types of organic material is a measure of the relative contribution of terrestrial carbon to a marine basin and an important factor in determining the basin's potential for oil or gas (Tissot and others 1974). Good photomicrographs of thin sections of spores and woody material are found in Bostick (1974) and in Staplin (1969). The reflectance of organic debris is also used as an indication of low-temperature metamorphism (Bostick, 1979).

h. *What can be learned from micropaleontology and palynology?* Many mudrocks and shales either contain few, if any, macrofossils or can only be studied in the subsurface from cuttings. Clearly, therefore, microfossils became essential. For example, they can be used to date or zone a thick shale section and thus to help unravel its stratigraphy and/or structure—in orogenic belts shales are the weakest lithology and, hence, favorite places for failure by overthrusting and complex folding. Even without complex structure, a proper knowledge of the microfauna can totally restructure our concept of a basin (Fig. 1.18). In addition, microfossils can also significantly contribute to our paleoecologic insight to a shale and even possibly help estimate distance from a shoreline. Consequently, most major petroleum companies use micropaleontologists and palynologists in their study of sedimentary basins, especially Mesozoic and younger basins. It is always best for the sedimentologist to work closely with the micropaleontologist or palynologist and to do so it is always helpful to have at least some understanding of their subjects.

Major groups of Phanerozoic microfossils include Foraminifera, ostracods, conodonts, algae, radiolarians, and all the palynormorphs—pollen, spores, acritarchs, chitinozoans, etc. An excellent overview of microbiostratigraphy is that of LeRoy (1977) and a recent overview of micropaleontology is Haq and Boersma (1978). The Foraminifera are probably the most studied. An early classic summary of how Foraminifera are useful is that of Phelger (1960). Pflum and Fredricks (1976) have elaborated and better documented how the Foraminifera provide ecologic information in the Gulf of Mexico, a largely muddy basin. For the conodonts, Lindström (1964) and Lindström and Ziegler (1972) are probably the best sources; for the algae, Wray (1977); for the ostracods, Neale (1967); Chaloner (1968) provides a short summary of the paleoecology of fossils spores; and D.J. Jones (1956) and Pokorny (1965) are two general texts. Another good source, which also provides many useful techniques for the study of shales in general, is the *Handbook of Paleontological Techniques,* by Kummel and Raup (1965). Although the foregoing is only a tiny sampling of the literature of this vast field, we suggest it in the hope of helping you get started.

i. *What is the significance of internal surface area?* For small particles such as those found in shales, the surface to volume ratio is relatively large and so surface effects become important for chemical and physical properties. For instance, release of natural gas to a well in shale may be a function of the accessible surface area of the rock. Work of this kind on shales is just beginning, but Thomas and Damberger (1976) have developed appropriate techniques using coals. One very useful result of their work is the demonstration that surface areas determined by the adsorption of CO_2 gas are higher than those determined using N_2 gas. This is explained by the inability of N_2 to penetrate pores smaller than about 5 Å. Thus a comparison of CO_2 and N_2 surface areas can be used as an indicator of pore-size distribution.

6. What Textural Parameters Should Be Measured and What Do They Mean?
As with composition, the study of the texture of shale is absolutely fundamental and yet, unfortunately, our understanding of how to

measure and interpret it lags far behind that of the mineralogy. Hence, one is on a petrographic frontier when one asks these questions, although it is clear that ultrathin thin sections of muds and shales would be very desirable (Martin, Litz and Huff 1979).

   a. *What proportions of clay, silt, and sand?* These proportions probably are best detemined in thin section for mudstones and shales, although some attempts have been made using X-ray diffraction (Till and Spears 1969). For muds the pipet and hydrometer methods (T. Allen 1974, pp. 194–200, 210) can be used, although even with modern muds, especially after they have dried, there is a question of what is being measured. When these ratios are mapped across a basin, its distal and proximal portions may be identified.

   b. *Percentage of large micas?* Large micas commonly are concentrated on bedding planes and commonly detrital. Somes shales have them in abundance, but in others they are absent. The hydraulic equivalent size of these flakes may well be a better measure of current competence than clay–silt–sand ratios obtained by artificial disaggregation.

   c. *Size ranges and modes of diverse detrital grains?* As with the proportions of clay, silt, and sand, these are best estimated in thin section. Size ranges and modes, especially of the silty shales, offer promise of refining inferences about provenance, an excellent early example being Folk's (1962) study of the Silurian Rochester and McKenzie Shales in the central Appalachians.

   d. *How well oriented are the different components of shale—the clay minerals and the silt and sand grains so commonly found in them—and what significance, if any, does their orientation have for paleocurrents?*

      i. *Perfection of framework orientation?* There is a direct relation between parting, and lamination and the perfection of clay mineral orientation of the framework. Blocky shales have randomly oriented clay particles, whereas laminated shales have clay particles well oriented parallel to the planes of lamination (Odum 1967, Moon 1972). The perfection of orientation is best observed and measured by X-ray diffraction. The degree of orientation appears to depend, first of all, on the amount of silt and sand present, and then on the clay mineralogy. Additional factors may be the water chemistry of the depositional basin as well as that of the early pore water, the organic compounds in solution, compactional history, and bioturbation. Clearly, orientation is a very complicated matter, one that is as yet not fully understood. However, it does seem that most of the clay mineral fabric of shales is of a post-depositional origin and, therefore, has little paleocurrent significance.

      ii. *Silt and sand grains?* Silt and sand grains in shales occur in three ways: as distinct thin laminae, as dispersed grains, and as isolated small domains or clumps. The thin laminae commonly imply a weak traction current, whereas dispersed grains generally imply either rapid deposition (dumping or flocculation), or bioturbation, or possibly wind-blown sand, and small clumps suggest a fecal pellet origin. All three types of occurrence are best seen in thin section. Paleocurrents can be inferred from the thin-section study of laminae of silt and sand and it may also be possible to study, with sections cut parallel to bedding, the shape orientation of dispersed grains in gritty shales using either the petrographic microscope or possibly SEM photographs.

e. *Burrowed and /or mottled textures?* These are very common in many argillaceous sediments and are almost always the result of biogenic activity. Burrows and mottles can range from distinct, identifiable structures in a sediment, with or without lamination, to totally stirred or churned mudstones (Fig. 2.9). They are commonly observed in thin-section and hand specimens; peels may be useful (Bull 1977). The amount of bioturbation can be estimated, at least semiquantitatively (Reineck 1963, Table 5), and depends on very high infauna populations as well as on sedimentation rates; i.e., where environmental factors are favorable, such as in a marine delta-front basin, bioturbation commonly is very abundant (in spite of high sedimentation rates). Usually slowly deposited sediments are the most thoroughly churned, but bioturbation can also be abundant when sedimentation rate is high under favorable environmental conditions—plenty of food, light, and oxygen (Beyers, 1977).

f. *Pore geometry—kinds and amounts?* This question can only be partially answered for modern muds—and hardly at all for ancient shales—most of the studies have been made by civil engineers for slope and foundation stability and by clay mineralogists interested in flocculation (Edzwald and O'Melia 1975). Pores are best studied by the SEM and by gas absorption. The pore system could also have a significant bearing on extracting gas from an organic-rich shale.

g. *Is there any significance to the shape of shale cuttings?* The manner in which a shale breaks in response to drilling reflects its physical properties. These properties could be useful clues in engineering studies of foundation stability just as we saw how the shape of cuttings was useful in stratigraphy (Fig. 2.5). However, type of drill bit, its weight, etc. also affect shape of the cuttings and hence should be considered.

h. *What can be seen by radiography?* This technique is fairly new for shales but is most informative, because it can show many things not readily seen by the naked eye, such as rootlets, fracture planes, and dessication structures, and can be used on modern muds as well as ancient shales (Fig. 2.10). Recognition of fracture planes and other inhomogeneities can be vital for engineering foundation studies and may be useful in understanding the response of a shale to fracturing and wire-line logging. Fraser and James (1969) provide exposure guides

TRACE FOSSILS   LITHOLOGY

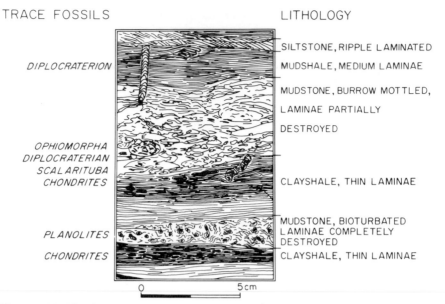

Figure 2.9 Mottling textures as seen in hand specimen.

for mud and shale and Ashley (1973) tells how to impregnate fine-grained sediments, such as Pleistocene varves, so they can be radiographed. A detailed general reference is Chapter 3 of Bouma (1969, pp. 140–244). Two good engineering applications, each with many illustrations, are Krinitzsky and Smith (1969) and Krinitzsky (1970). Other examples include Hester and Pryor (1972) and Zimmerman (1972). Radiography is almost always used for the study of modern deep-sea sediments (Hesse 1977). Today, radiography should be considered an absolutely standard technique for the study of shales.

7. What Name Is To Be Used and How Should the Shale Be Classified? After the previous questions have been answered enough information will be on hand to attempt answers to these. The foregoing laboratory analyses will chiefly supply semiquantitative or quantitative mineralogic

Figure 2.10 Radiography is very effective for the study of mud and shale,→ because it reveals macroscopic details not apparent to the eye alone: A (top facing page) Lacustrine mud from the lower valley of the Mississippi River (Krinitzsky and Smith, 1969, Fig. 21) actually has a complex system of dessication cracks, and B: Radiograph of Devonian New Albany Shale from western shelf of Illinois Basin (lower left) shows completely bioturbated shale with pyrite filled burrows (pb) whereas radiograph from New Albany Shale in deep, poorly oxygenated central part of basin (lower right) shows excellent lamination with some pyrite nodules and a carbonate lens (lamination is preserved because bottom conditions were too toxic). Photographs courtesy of R.M. Cluff, Illinois Geological Survey, Urbana, Illinois.

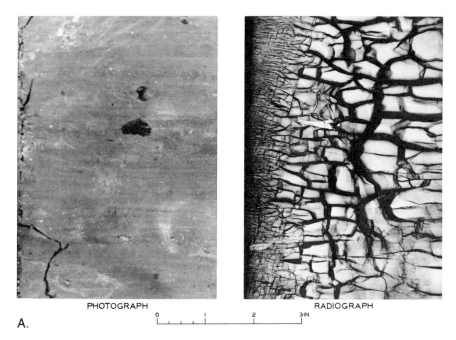

PHOTOGRAPH   RADIOGRAPH

A.

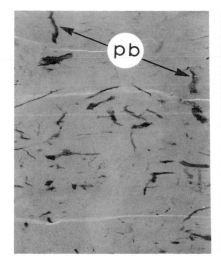
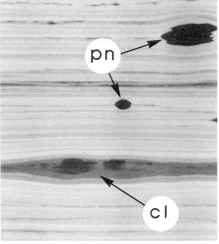

B.

Scott W. Starratt
Dept. of Paleontology
U. C. Berkeley
Berkeley, Ca. 94720

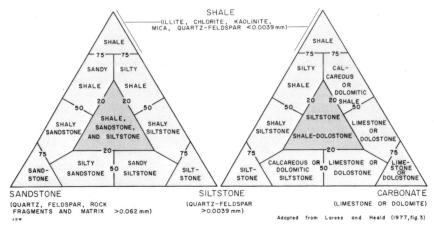

Figure 2.11 Triangular diagram of shale and its commonly associated lithologies (Adopted from Larese and Heald, 1977, Fig. 3).

and textural data—the proportions of the different mineral phases present and the relative proportions of sand, silt, and clay. Therefore, precise mineralogic and/or textural classifications are now possible; good examples are Larese and Heald (1977) and Lewan (1978). A triangular diagram still has utility, especially when only mineralogy is considered (Fig. 2.11). It is also possible that thin sections and radiography may alter our perception of bedding as seen in the field, i.e., show the presence of lamination and/or biogenic structures that have escaped us in the field. From still another viewpoint, it is also possible that we might make a specialized classification based on geochemical study using color, precent total carbon (Fig. 2.12) and grain size as assessed by the percentage of silt (see also Table 1.2).

8. What Is the Best Way to Organize Petrology and Texture into a Useful, Concise, and Coherent Petrographic Report? Although each investigator will probably wish to devise his or her own form, we have included one that we have found useful (Table 2.7). The table by Kemper and Zimmerle is a good example of how a summary petrographic table might be organized (Table 2.8). As in all petrology, making systematic observations is essential for success and, therefore, a petrographic form is usually necessary. Photomicrographs are usually essential, although the technology of how to obtain informative black and white or even color ones of shales is still to be fully developed. We have included colored photomicrographs of all major classes of shales and several metamorphic equivalents (Fig. 2.13, see insert between pages 118 and 119). Pen and ink drawings of photomicrographs should not be discounted.

9. What Can Be Learned from the Study of Inorganic Chemistry? Making chemical analyses was, most probably, the first thing that was ever done to a shale in the laboratory long ago in the early nineteenth

Organic Geochemistry of Phosphoria Black Shale

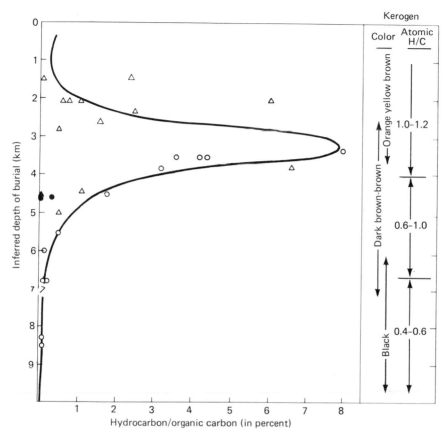

Figure 2.12 Hydrocarbon-to-organic carbon ratio and composition of kerogen plotted against inferred maximum depth of burial for the Phosphoria Formation (Claypool and others, 1978, Fig. 7).

century, if not before. Today, there are a number of elementary texts, such as those by Brownfield (1977) and Krauskopf (1979), with material on the inorganic chemistry of sediments. The common chemical techniques of today are summarized in Table 2.9.

a. *The major elements?* So far, not much has come from such study because the analyses have not been related to the history of particular basins. If they were, it might be possible to interpret the chemistry of the shales using evidence from their interbedded sandstones and carbonates, which we understand much better. However, some pioneering studies of the bulk chemistry of shales have been made by Nanz (1953), Ronov and Migdisov (1971), and Björlykke (1974).

b. *The trace elements?* Geochemists have long been fascinated by trace elements as criteria for distinguishing marine from nonmarine muds

Table 2.7. Shale Petrology Report Form*

Name: _____ DATE: _____	

Formation: _____

Rock Type: _____ Alteration: _____

TS orientation: _____ vertical _____ horiz.   Type: _____ polished _____ covered

Photographs:_____

****

Megascopic description: Include color, lamination (absence or presence and perfection thereof), bioturbation, and precentage of siltstone, carbonate, and fossil debris, including kinds, abundance, and orientation. Fractures, kinds and quantity? Short but useful description for field or core study.

Larger than 2 μm fraction: Size, shape, sorting, orientation, dispersal and diagenesis.

   Quartz:
   Carbonate:
   Pyrite:
   Other:

Dispersed organic matter: Types, size, amount, orientation, dispersal, alteration.

   Coaly particles:
   Sapropelic or amorphous:
   Microfossils and spores:
   Fluorescence:

Clay: Orientation, birefringence, remarks

Microsedimentary strutures: Physical and biogenic

Scanning electron microscope description:

Chemical characteristics:

Petrographic type and significance: Give microfacies, if possible, and interpretation of depositional and postdepositional history insofar as is possible. Paragenesis diagram may be helpful. Relate microfacies to position in basin or to local vertical sequence and distance from founding lithologies.

Microfacies interpretation and remarks:

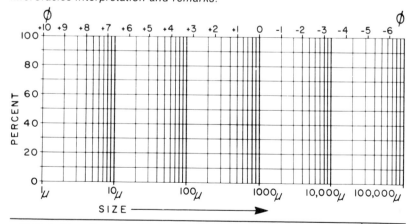

Adapted from the work of Richard D. Harvey and R.M. Cluff, Illinois Geological Survey, Urbana, Illinois, with their kind permission.

Table 2.8. Comparison of Fischschiefer (Cretaceous) of Rethmar and Helgoland (Translated from Kemper and Zimmerle, 1978, Table 1)

Components/Parameters		Mittelland Kanal, Rethmar	Düne, Helgoland
Microscopic Appearance		Pure calcite laminae rich in coccoliths interlaminated with dark, organic-rich clay	Clean calcite lenses and masses rich in coccoliths interlaminated with dark, organic-rich clay
Mineral Composition (X-ray)		CALCITE—quartz, pyrite, muscovite/illite, kaolinite and siderite; leached fraction smaller than 20 μm as above, but contains montmorillonite	CALCITE—quartz, pyrite, muscovite/illite and kaolinite; leached fraction smaller than 20 μm as above, but also contains natrolite(?), chlorite(?) and montmorillonite
Coarse Silt and Fine Sand ($>$35 μm)	Light minerals	SANIDINE, albite, quartz and glauconite	QUARTZ $>$ feldspar (sanidine) $>$ microcline $>$ plagioclase) and glauconite
	Heavy minerals ($>$2.9)	PYRITE + FRAMBOIDALPYRITE, brownish collophane, biotite, apatite, sphalerite, chlorite, siderite	FRAMBOIDALPYRITE, brownish collophane, anomolous zircon-monazite, siderite, rutile, sphalerite, clinozoisite, tourmaline, barite, anatase, brookite, and garnet plus single grains of volcanic biotite, apatite and titanite
Organic Materials		Type: bituminous, sapropelitic (kerogen I according to Tissot) Abundance: generally about 5% Thermal maturation: mature, beginning of oil generation Oil source bed potential: good	Type: bituminous, sapropelitic (kerogen I according to Tissot) Abundance: generally about 10% Thermal maturation: maturation: mature, beginning of oil generation Oil source bed potential: very good

Table continued on page 116

Table 2.8. (*Continued.*) Comparison of Fischschiefer (Cretaceous) of Rethmar and Helgoland (Translated from Kemper and Zimmerle 1978, Table 1)

Components/Parameters	Mittelland Kanal, Rethmar	Düne, Helgoland
Fossil Content	Dominantly nectoplanktonic forms with only minor Hedbergellian forms	Dominantly nectoplanktonic forms, Hedbergellian forms somewhat more common than in Rethmar
Porosity (%)	31.4 to 32.3	34.5 to 35.5
Permeability (millidarcys)	0.08 to 3.4	—

Table 2.9. Common Chemical Techniques and Their Utility

Technique	Utility
Field chemistry Use dilute HCl and/or the tests for dolomite (Friedman 1959) on cuttings or in the outcrop	Effervesence indicates presence calcite and lack of effervesence and pink color indicates dolomite
Laboratory chemistry Bulk chemical analysis Use atomic absorption or X-ray fluorescence	Usually about 8 to 12 oxides are reported. Bulk chemical analyses may be of economic interest (how much alumina or potash is in the shale) and can define a maturity index for shales (Björlykke 1974, p. 263)
Trace elements Same techniques as above plus neutron activation analysis	Helpful in mineral exploration, possibly useful in determining environment of deposition, and have even been used to correlate shales
Microprobe analysis Can be done quantitatively with the electron microprobe or semi-quantitatively with an energy dispersive attachment to a scanning electron microscope	Permits determination of the chemistry of separate grains as small as 1 μm; much more promising than bulk chemistry, because different particles usually have had a different history (i.e., sedimentary rocks are not equilibrium mineral assemblages as are metamorphic rocks and most igneous ones)
Elemental Analysis of Organic Matter A CHN analyzer	Percentage carbon appears to be an important variable in controlling other chemical and physical properties of a shale; additionally, the H/C and N/C ratios may reflect diagenetic history

and shales, chiefly because trace elements can be analyzed quickly using such techniques as X-ray fluorescence (Shimp and others 1969). Reeves and Brooks (1978) provide the most complete recent review of the methods of trace element determination and also give information on geological applications and statistical analysis. However, in spite of many studies, there is still no accepted threshold of single or combinations of trace elements that discriminates between marine and non-marine, although boron and vanadium appear to have produced the best results so far.

c. *Exchangeable cations?* Too often neglected in chemical analyses, exchangeable cations can potentially provide much useful information. For instance, in bulk chemical analyses the exchangeable ions need to be distinguished from fixed ions before any type of provenance interpretation is made. Spears (1973) has also suggested that exchangeable cations can be used in paleosalinity studies, high Mg^{2+}/Ca^{2+} ratios indicating marine deposition. However, calcite interferes with the measurements and groundwater movement during early diagenesis may also have a strong effect. Exchangeable cations appear to have at least one important industrial application—determining how a shale may react to a drilling fluid (O'Brien and Chenevert 1973).

   d. *Pore-water chemistry?* The chemistry of the pore fluid can provide information about mineral transformations. High silica in interstitial waters has been used as evidence of the dissolution of diatoms (Heath 1974), and depletion of Mg^{2+} with depth as evidence of uptake by carbonates (Lerman 1975) and of clay mineral transformation in deep burial in oil fields (Burst 1976). Another approach is to contrast the pore water of shales to that of their associated standstones and thus to better explain the history of secondary mineral transformation in both (Schmidt 1973).

   e. *Stable isotope geochemistry?* This is most informative when studied on specific substances; for example $^{13}C/^{12}C$ on kerogen and on the carbonate minerals; $^{18}O/^{16}O$ on the carbonates and possibly on the silica polymorphs; and $^{34}S/^{32}S$ on pyrite and gypsum. The best approach is to contrast the isotopes of an element which occurs in two different minerals in the same formation and thus to learn more about the paragenetic history of the shale. For instance, $^{13}C/^{12}C$ on siderite, calcite, and dolomite from the same unit can help differentiate the stage of development of each because the carbon source may change during diagenesis (Hudson 1977).

   Although expensive, a few properly chosen isotopic determinations can go a long way. Most promising is the study of oxygen for paleo-temperatures (Murata and others 1977, Eslinger and others 1979) and sulfur; the latter appears to be related to sedimentation rate (Goldhaber and Kaplan 1975).

10. Organic Chemistry? The study of organic compounds in sediments began in shales as long ago as the 1920s, primarily because shales were believed to be the source of most of the petroleum found in sandstones and carbonates. Two early classic studies are those by Trask and Wu (1930) and Treibs (1934). Today, organic geochemistry contributes to the understanding of the thermal history of a shale and also enhances our knowledge of the types of organisms that have produced the original organic matter—a kind of paleobiology. By and large, the paleobiochemical indicators (biologic molecules, such as amino acids, phytane, or porphyrins) are more difficult to study and require more expensive and sophisticated equipment than do the thermal indicators.

   a. *What are the best indicators of thermal history?* The easiest indicator to use is the color of kerogen in transmitted light (Burgess 1974), where

Figure 2.13. Twenty-two photomicrographs illustrate the shale types of Table 1.2, selected sedimentary structures found in thin sections of shale, several famous shales; and some of the metamorphic equivlents of shale. Without pictures such as these it is difficult to convey to the reader all the fascinating features that can be seen in thin sections of shale. Colored photomicrographs best illustrate most of these features. All photomicrographs are by Wayne A. Pryor using a Leitz Photomicroscope II. M. Lewan (AMOCO Research Laboratory) furnished (f), (p), (q) and (s).

(a) Siltstone: argillaceous with interstitial clay and discrete lenses of illitic clay, bioturbated. Ordovician, St. Peter Sandstone, S. Illinois, U.S.A.

(b) Mudstone: calcareous and quartzose with lense-shaped, dark organic-rich filled burrow. Ordovician, Fairview fm., Cincinnati, Ohio, U.S.A.

(c) Mudstone: red, calcareous and quartzose. Calcite occurs as veinlets. U. Triassic, Rote Wand-Keuper, Heilbronn, Germany.

(d) Claystone: calcareous and very uniform with pelecypod fragments. U. Jurassic, U. Kimmeridge-Pavlovia zone, Dorset Coast, England.

(e) Claystone: montomorillonitic (smectitic) with well defined wisps of kerogen. Eocene, Tarras Clay, Fehmarn Island, Schleswig-Holstein, Germany.

(f) Siltstone: ripple laminated and quartzose. Pleistocene. Intertidal facies, San Felipe, Baja California, Mexico.

(g) Mudshale: quartzose, burrowed and bioturbated with well defined silt laminae. Cambrian, Mt. Simon fm, Dugape Co., Illinois, U.S.A.

(h) Clayshale: calcareous with kerogen-rich laminae, quartz grains, and horizontal lamination. Eocene, Green River fm-Laney mbr, Washakie Basin, Wyoming, U.S.A.

(i) Clayshale: kerogen-rich and calcareous. Horizontal laminae defined by differences in composition. L. Cretaceous, Aptian Fischschiefer fm., Rethmar, Germany.

(j) Quartz argillite: chloritic, felspathic with graded laminae. Chlorite forms interstitial matrix and defines some laminae. Precambrian-Cobalt, Gowganda fm., Minnesota, U.S.A.

(k) Quartz slate: chlorite matrix and biotite. Incipient slip cleavage at left. M. Triassic, Quartenschiefer, Lukmanier Pass, Switzerland.

(l) Phyllite: quartzose and sercitic with biotite, magnetite porphyroblasts, and cross laminae. Cambrian, L. Devillian-Blaumont series, Opperbais Quarry, Belgium.

(m) "Oil Shale": kerogen-rich, quartzose, calcareous and dolomitic *mudshale*. Yellow laminae are kerogen. Loop structure (Fig. 16) and microfaulting (in center of photograph). Eocene, Green River fm.-Laney mbr. Washakie Basin, Wyoming, U.S.A.

(n) "Oil Shale": fossiliferous, calcitic *mudstone* with kerogen, contains abundant pelecypod, ostracod and brachiopod fragments. L. Carboniferous, Visean-Namurian, Cawdor Quarry, Sheffield, England.

(o) "Oil Shale": kerogen-rich, dolomitic siltstone. Kerogen interlaminated with lenticular blebs of silt size dolomite crystals. Some quarts and illite. Lenticular lamination. L. Permian, Assise de Muse fm., Autun, France.

(p) Mudshale: calcitic with strong compositional and color lamination. Miocene, Monterey Shale, Elwood Beach, Santa Barbara Co., California, U.S.A.

(q) Mudstone: quartzose and fossiliferous mudstone. Note small chert nodule (blue) and pyrite filled foraminifer. Crossed nicols. Eocene, Kreyenhagen fm., Garza Creek, Kings Co., California, U.S.A.

(r) Mudshale: quartzose and limonitic with ripple laminae of quartz silt. Very well laminated. U. Cambrian, Alum Shale, Southern Sweden.

(s) Claystone-Mudshale-Siltstone: interlaminated claystone, quartzose mudshale and quartzose siltstone. Shows how finely interlaminated the different shale types can be. Good example of textural, color and compositional lamination. Precambrian, Nonesuch Shale, Big Iron River, Ontonagon Co., Michigan, U.S.A.

(t) Pelletal Mudstone: argillaceous with calcite cement. A pelletal carbonate rock with shale characteristics. Eocene, Green River fm., Laney mbr., Washakie Basin, Wyoming, U.S.A.

(u) Mudshale: quartzose with calcite crystals (stained pink) that replaced halite. Eocene, Uinta fm., Saline Facies, Uinta Basin, Utah, U.S.A.

(v) Clayshale: quartzose with kerogen. Phosphatic nodule (brown-orange) and deformed laminae. L. Carboniferous, Oil Shale Gp., Fells Seam, Westwood Pit, Edinburgh, Scotland.

(w) Mudstone: dolomitic and kerogen-rich with abundant quartz. Bioturbation has destroyed all primary lamination. U. Permian, Kupferschiefer, Cornberg Quarry, Werra District, N. Hessen, Germany.

(x) Siltstone: kerogen-rich and argillaceous with very well defined horizontal and vertical burrow mottling. U. Devonian, Brallier fm., Appalachian Basin, Virginia, U.S.A.

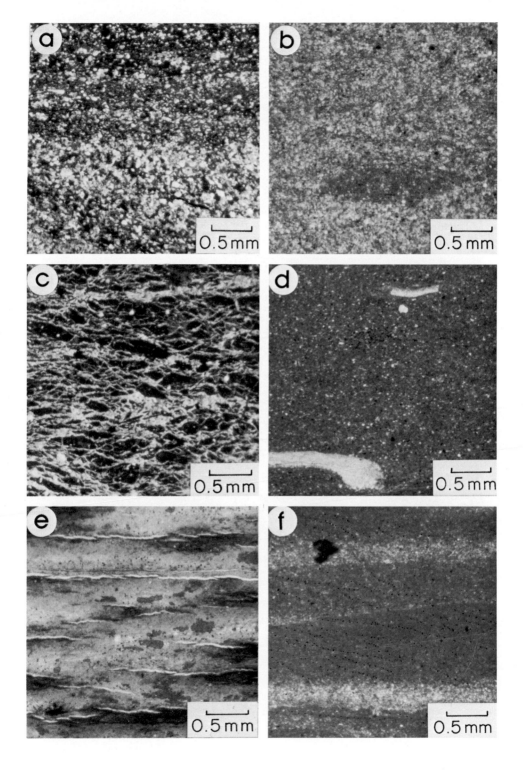

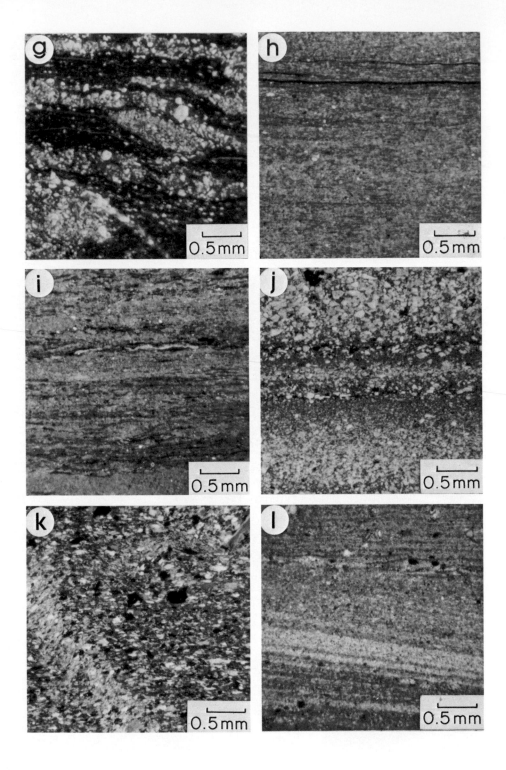

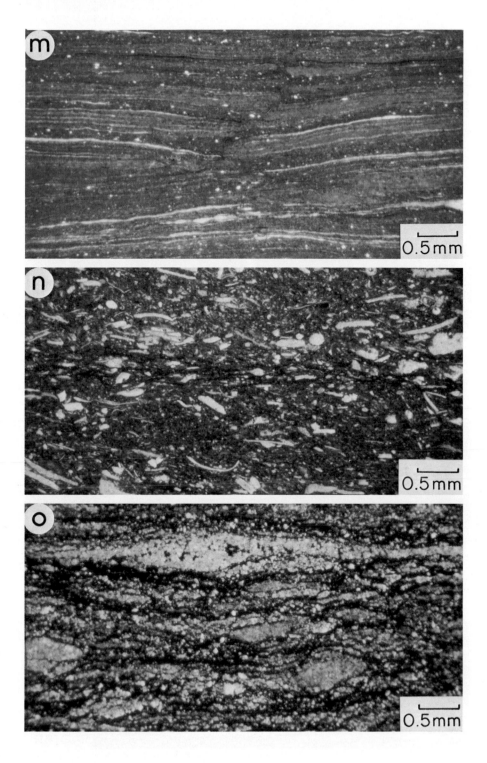

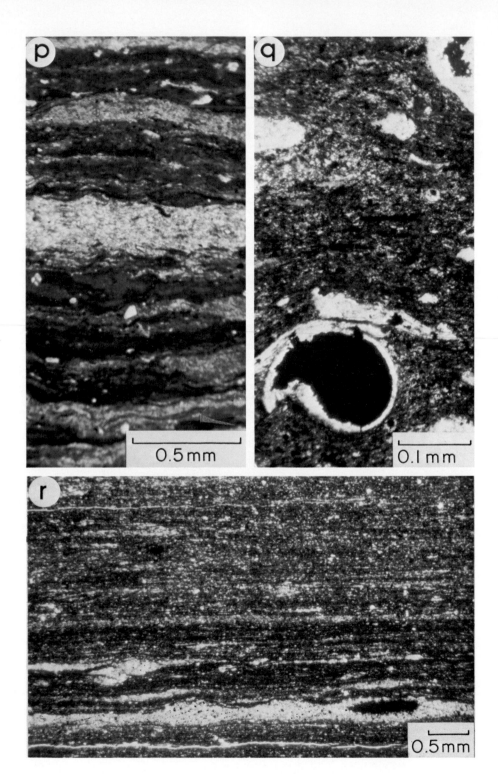

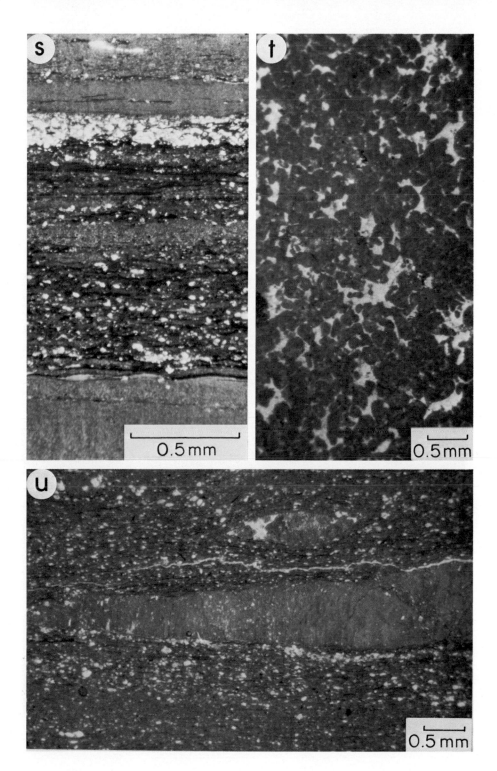

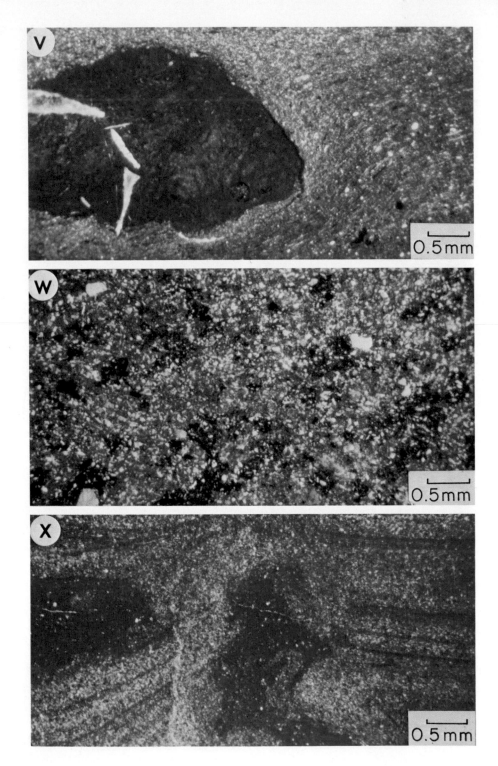

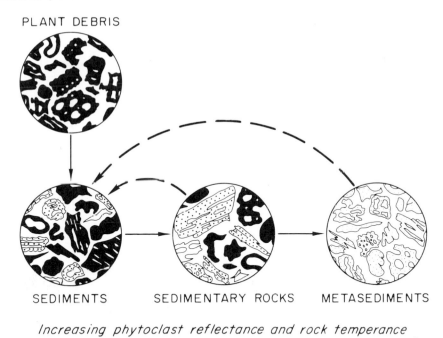

PLANT DEBRIS

SEDIMENTS   SEDIMENTARY ROCKS   METASEDIMENTS

Increasing phytoclast reflectance and rock temperance

Figure 2.14 Reflectance of phytoclasts and temperature. (Redrawn from Bostick, 1974, Fig. 6).

opaque kerogen corresponds to high temperature and translucent to low. A related technique is the color of conodonts (Epstein and others 1977). Still another that is fairly simple is the proportion of organic carbon that can be extracted with such solvents as chloroform. This method, however, should be used with some care, because the ratio attains a maximum with increasing burial depth (temperature) and then declines (Fig. 2.12). Another standard technique is the use of carbon–hydrogen–oxygen analysis of kerogen (Tissot and others 1974). One can also use the reflectance of small woody particles, called phytoclasts (Fig. 2.14), and the sharpness of the illite peak (Fig. 2.15). The recent review by Bostick (1979) is good to begin the study of reflectance of phytoclasts. In metamorphic sediments one can examine the sharpness of the graphite peak on X-ray tracings (Grew 1974).

b. *What does the study of paleobiochemical indicators tell us?* Two problems that have been studied using biologic molecules are the development of life in the Precambrian (Schopf and others 1968, Anders and others 1973) and the origin of the carbon in a depositional basin—is it marine or terrestrial? Two promising indicators of the source of the C are the $^{13}C/^{12}C$ ratio (Newman and others 1973) and phenolic aldehydes (Gardner and Menzel 1974).

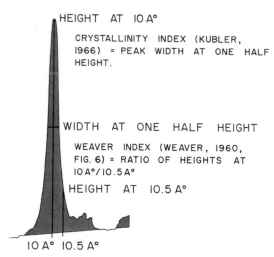

HEIGHT AT IO A°

CRYSTALLINITY INDEX (KUBLER, 1966) = PEAK WIDTH AT ONE HALF HEIGHT.

WIDTH AT ONE HALF HEIGHT

WEAVER INDEX (WEAVER, 1960, FIG. 6) = RATIO OF HEIGHTS AT IO A°/ IO.5 A°

HEIGHT AT IO.5 A°

IO A° IO.5 A°

Figure 2.15 Sharpness of the illite peak and two ways to measure it.

MAKING A SYNTHESIS: DEPOSITIONAL ENVIRONMENT AND BASIN ANALYSIS

Consideration of the shale and its lateral equivalents over an entire basin—be the basin one that covered much of an old craton, an entire continental margin, or even a small ocean, such as the present Arctic Ocean—is the only way the sedimentologist can "put all the parts together" into an integrated whole. Such a synthesis offers the potential of relating the smallest details of, say, petrology and texture, on the one hand, to the distribution of major mud or shale facies across a basin and thus exploring how facies and basin-wide bathymetry and paleocirculation are all related. To achieve this ideal, a *systematic inventory of facts* is, again, essential.

Systematic mapping, mostly using wire-line logs and geophysics, and especially seismic stratigraphy (Payton 1977), supplemented by the careful study of available cores and outcrops, is the means by which most of the "regional" aspects of a shale—its basin-wide geometry plus its facies types and their distribution—are obtained. Systematic mapping defines both the external geometry and the internal organization of the *depositional sequences* within the mudrock or shale, a depositional sequence being nothing more than "a stratigraphic unit composed of relatively conformable successions of genetically related strata and bounded at its top and base by unconformities or correlative unconformities" (Mitchum and others 1977, Fig. 1). A depositional sequence may range in thickness from a few tens of meters to hundreds or even thousands of meters. Such a sequence is a natural sedimentary packet or bundle that gives much basic insight into the shale. The environments of deposition of such bundles always should be determined, at present a very major task. Criteria that are helpful and can

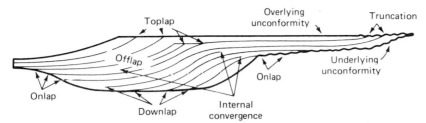

Figure 2.16 Terminology for depositional sequences (Mitchum and others, 1977, Fig. 1). Published by permission of the authors and the American Association of Petroleum Geologists.

be mapped, include: silt content, color, trace elements and clay mineralogy (Weaver, 1978, p. 151–164), but fossils, if present, are the most effective. The geometry and terminology of these bundles have recently been standardized and well illustrated (Fig. 2.16). Shale bundles may be separated from one another by subtle breaks and hiatuses in sedimentation (Baird 1976)—some of which are expressed by concentrations of concretions (Voight 1968)—as well as by density contrasts and bedding that can be seen on the scale of seismic sections. Seismic sections are very helpful in detecting slight discordances within shaly sequences and can also be used in basins where few logs are yet available. Moreover, shale petrology and geochemistry become most meaningful—become *stratigraphic petrology and geochemistry*—only when fully integrated with the depositional sequence (Fig. 2.17).

Cross sections and maps are essential for a basin analysis. Cross sections should be built parallel and perpendicular to depositional strike and across the tectonic grain of the basin or across its major structures as well. Maps should be made of what you consider to be the most important stratigraphic units and of specific properties, such as color, fossil content, carbon isotopes, thickness, sand–shale ratio, etc. In short, whatever you have systematically recorded and studied has the potential to become the building block for a sedimentologic, geochemical, or paleoecologic map of some kind.

To bring all this information together, some kind of report form will be needed. Whatever the final report—be it a few or many pages—it is always helpful to both you and the reader to condense your results into a useful summary table of some kind, of which the organization of Table 2.10 is but one possibility. Depending on the problem at hand, a very different table may suit your needs. As a supplement to necessary maps and cross sections, however, a table is essential.

Because our knowledge of shales lags so far behind that of sandstones and carbonates, there are very few studies of shaly basins which we can use as specific models and as yet no general "mud models" for mudrocks and shales. Hence, our methodology comes chiefly from the study of sandy and carbonate basins, some general sources being Selley (1976, pp. 322–355),

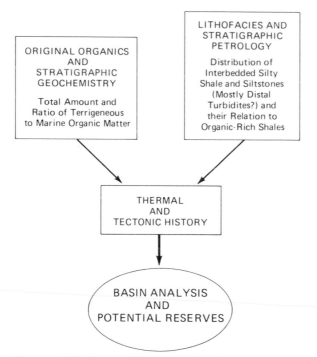

Figure 2.17 General flow chart for development of the hydrocarbon potential of a shaly basin. The first step is nearly always to carefully delineate the many minor stratigraphic units present in shaly basins and to determine their genetic relationships to one another—to do sedimentology first.

Wilson (1975, pp. 348–379), Potter and Pettijohn (1977, pp. 340–363) and Reading (1978). Another useful source of information is the collection of papers found in Volume 1 of Société Nationale Elf-Aquitaine (Production) for 1977.

In making an analysis of a shaly basin, there are four recurring questions that virtually always need answers. Although these questions are not explicit in Table 2.10, you will always need to answer them for your final synthesis and analysis. Fortunately, most of the data are already at hand and you only need to return to the first two sections of the Question Set to obtain it. So, if you have done your work well and are fortunate to have enough data, your task should not be too hard—and it should be of great fun and interest as well! All four questions are very process oriented and deserve your full attention.

1. Where Did the Mud Come From? This question is fundamental, because to answer it you need to know something about the *paleogeography* of your basin—what was depositional strike, how far away was the source of the mud, and what kind of terrain has provided it? Have the transport paths of mud and sand always been the same (Fig.

Table 2.10. Format for the Analysis of Shaly Basins

Properties	Methods
Time span Stratigraphic range and possible worldwide megasequence	Paleontology, especially micropaleontology in the Phanerozoic; in the Precambrian primarily radioactive dating
Geometry and basin type Shape, area, volume, and maximum thickness	Mostly subsurface mapping plus possible seismic and gravity study
Sedimentology Lithic fill: Kinds, proportions, and distribution of major depositional environments and system (implying kinds, proportions, and distribution of major shale facies and interbedded and neighboring lithologies)	Outcrop and subsurface lithologic correlation plus paleontology, especially micropaleontology, as well as systematic environmental analysis from both outcrop and subsurface; possible use of seismic stratigraphy
Composition: Organic geochemistry, mineralogy, texture and petrographic types of shale, silt, and related lithologies	Mostly thin-section and binocular petrology plus some X-ray identification
Paleocurrents: Integrate depositional systems and help predict major sedimentary trends; oxygenation and nature of water masses in basin are also potentially discernible	Primarily the systematic mapping of cross-beds and sole marks in interbedded siltstones, sandstones, and possibly in carbonates, supplemented by facies distribution and biological and mineral provinces plus organic geochemical facies
Structural style Predepositional: May be hard to assess but commonly the chief control on basin geometry and regional paleoslope	Seismic and gravity studies plus subsurface and regional geology
Syndepositional: Abundance and magnitude of regional arches, hinge lines, and/or growth faults plus unconformities	Subsurface stratigraphy, abundance and magnitude of regional and local unconformities, plus seismic sections
Postdepositional: May consist of major fold systems and/or thrusting or chiefly be normal faulting, or the basin may be essentially undeformed with only broad regional arches; shales commonly associated with major overthrusts	As above plus field mapping

Table 2.10. *Continued*

Properties	Methods

Tectonic setting
  Summarizes basin type and may
    indicate relationship to possible
    plate tectonic schemes
	Based on nature of lithic fill, structural style, and relation to either present or inferred continental margin; sandstone composition can be helpful in assessing possible plate tectonic setting in pre-Mesozoic basins; as yet little is known about shale composition and plate tectonics

Paleoclimate
  Identification of broad world-
    wide climatic zones
	Climatically sensitive sediments (coals, evaporites, diamictites, and carbonates) and fossils, plus evidence from continental drift and paleomagnetics

Thermal burial history Closely related to economic potential of basin and depends on burial depth and geothermal gradient and thus ultimately on tectonic history	Organic and inorganic evidence, the former including coal rank, phytoclasts, plus kerogen and conodont color, and the latter including crystallinity of illite, percentage expandables in illite, zeolites, and metamorphic mineral assemblages
Economic Depends on objective but nearly always involves volume and distribution of host sediment in combination with local and/or regional structure; source rock potential for petroleum, oil and gas shales, and some heavy metals (uranium) are the chief economic interests	All of above plus geochemical study of organic content

2.18)? Although it may strike you as bizarre, it is often as hard to answer this question in the Recent—say, a muddy, shoaling estuary—as it is in an ancient shaly basin.

2. How Was the Mud Transported to its Final Depositional Site? This question, like most of these four, quite possibly will only be answered at the very end of your investigation. For example, was transport by pelagic suspension or as the weak, decaying tail of a distal turbidity current, or possibly even as eolian dust?

3. At What Water Depth, Sedimentation Rate, Oxygenation, and Toxicity to Life Was the Mud Deposited? To answer this you will need to sort out and recreate to your fullest the original depositional environment using every means available to you from the first two sections of the Question Set.

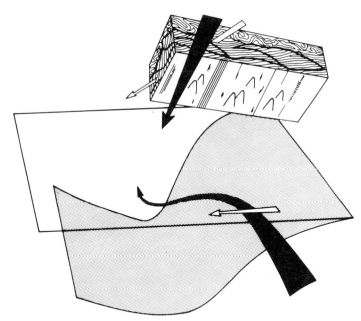

Figure 2.18 Divergent currents in elongate turbidite basin. Thin bed with solemarks and ripple bedding shows divergent sediment transport paths within single turbidite sandstone—the solemarks down the basin axis and the ripples down the basin slope. (Redrawn from Seilacher and Meischner, 1964, Fig. 10). How general, if at all, is this condition?

4. What Has Happened to the Mud Since Its Deposition? The answer involves the *total postdepositional history* of the mud, which includes such aspects as its compactional history (which in some basins may have been very, very complex, with several periods of uplift, erosion, and unconformities), pore-water chemistry, possible mineralogic transformations, and even the development and magnitude of fractures and microfractures plus folds and faults.

So now, with the help of the Annotated and Illustrated Bibliography, *it is all up to you.* GOOD LUCK!

References

Ager, D. 1963. Principles of Paleoecology. McGraw-Hill Book Co., New York, 371 pp.

Allen, J.R.L. 1964. The Nigerian continental margin: Bottom sediments, submarine morphology and geological evolution: Mar. Geol., v. 1, p. 289–332.

————.1968. Current Ripples. North Holland Publishing Co., Amsterdam, 433 pp.

Allen, T. 1974. Particle Size Measurement. Chapman and Hall, London, 454 pp.

Allersma, E. 1971. Mud on the oceanic shelf off Guyana. In: Symposium on Investigation and Resources of the Carribean Sea and Adjacent Regions. Willemstad, Curaçao, 10–26 November 1968, FAO/UNESCO Paris p. 193–203.

Alling, H.L. 1945. Use of microlithologies as illustrated by some New York sedimentary rocks: Geol. Soc. Amer. Bull., v. 56, p. 737–756.

Anders, E.R., R. Hayatsu, and M.R. Studier. 1973. Organic compounds in meteorites: Science, v. 182, p. 781–790.

Angelucci, A., E. de Rosa, G. Fierro, M. Gnaccolini, G.B. LaMonica, B. Martinis, G.C. Parea, T. Pescatore, A. Rizzini, and F.C. Wezel. 1967. Sedimentological characteristics of some Italian turbidites: Geol. Romana, v. 1, p. 345–420.

Armstrong, R.L. 1968. Sevier orogenic belt in Nevada and Utah: Geol. Soc. Amer. Bull., v. 79, p. 429–458.

Aronson, J.L., and J. Hower. 1976. Mechanism of burial metamorphism of argillaceous sediment: 2. Radiogenic argon evidence: Geol. Soc. Amer. Bull., v. 87, p. 725–744.

Ashley, G.M. 1973. Impregnation of fine grained sediments with a polyester resin: A modification of Altemuller's method: Jour. Sed. Petrol., v. 43, p. 298–301.

Asmus, H.E. 1975. Controle estructural da deposicâo Mesozoica nas bacias da margem continental Brasileira: Rev. Brasliera Geocien., v. 5, p. 160–175.

Asquith, D.O. 1970. Depositional topography and major marine environments, Late Cretaceous, Wyoming: Amer. Assoc. Petrol. Geol. Bull., v. 54, p. 1184–1224.

Atherton, E., G.H. Emrich, H.D. Glass, P.E. Potter, and D.H. Swann. 1960. Differentiation of Caseyville (Pennsylvanian) and Chester (Mississippian) Sediments in the Illinois Basin. Illinois State Geol. Survey Circ. 306, 36 pp.

Baird, G. 1976. Coral encrusted concretions: A key to recognition of a "shale on shale" erosion surface: Lethaia, v. 9, p. 293–302.

Barker, C. 1972. Aquathermal pressuring role of temperature in development of abnormal-pressure zones: Amer. Assoc. Petrol. Geol. Bull., v. 56, p. 2068–2071.

Beckmann, H. 1976. Geological Prospecting of Petroleum. John Wiley and Sons, New York, 183 pp.

Berkman, D.A., and W.R. Ryall. 1976. Field Geolgoists' Manual. Australian Inst. Mining and Metallurgy, Monograph Ser. No. 9, 295 pp.

Berner, R.A. 1968. Calcium carbonate concretions formed by the decomposition of organic matter: Science, v. 159, p. 195–197.

Berner, R.A. 1971. Principles of Chemical Sedimentology. McGraw-Hill Book Co., New York, 240 pp.

————. 1978. Sulfate reduction and the rate of deposition of marine sediments. Earth Planetary Science Letters, v. 37, p. 492–498.

Biscaye, P.E. 1965. Mineralogy of sedimentation of recent deep-sea clay in the Atlantic Ocean and adjacent seas and oceans. Geol. Soc. Amer. Bull., v. 76, p. 803–832.

Blissenbach, E. 1954. Geology of alluvial fans in semi-arid regions: Geol. Soc. Amer. Bull., v. 65, p. 175–190.

Björlykke, K. 1974. Geochemical and mineralogical influence of Ordovician island arcs on epicontinental clastic sedimentation. A study of Lower Palaeozoic sedimentation in the Oslo region, Norway: Sedimentology, v. 21, p. 251–272.

Blatt, Harvey, Gerard Middleton, and Raymond Murray. 1980. Origin of Sedimentary Rocks, 2nd ed. Prentice-Hall, Inc., Englewood Cliffs, N.J., 782 pp.

Borradaile, G.J. 1977. Compaction estimate limits from sand dike orientations: Jour. Sed. Petrol., v. 47, p. 1598–1601.

Bostick, N.H. 1974. Phytoclasts as indicators of thermal metamorphism, Franciscan assemblage and Great Valley Sequence (upper Mesozoic) California. In: R.R. Dutcher, P.A. Hacquebard, J.M. Simon, Eds. Carboniferous materials as indicators of metamorphism. Geol. Soc. Amer. Spec. Paper 153, p. 1–18.

———. 1979. Microscopic measurement of the level of catagenesis of solid organic matter in sedimentary rocks to aid exploration for petroleum and to determine former burial temperatures—A review. In: P.A. Scholle and P.R. Schluger, Eds. Soc. Econ. Paleont. Mineral. Sp. Pub. 26, p. 17–43.

Bouma, A.H. 1969. Methods for the Study of Sedimentary Structures. John Wiley-Interscience, New York, 446 pp.

Bradley, W.H. 1931. Non-glacial marine varves: Amer. Jour. Sci., Ser. 5, v. 22, p. 318–330.

Brown, L.F. Jr., S.W. Bailey, L.M. Cline, and J.S. Lister. 1977. Clay mineralogy in relation to deltaic sedimentation patterns of Desmonian cyclothems in Iowa–Missouri: Clays Clay Minerals, v. 25, p. 171–186.

Brownlow, A.H. 1977. Geochemistry. Prentice-Hall, Inc., Englewood Cliffs, New Jersey, 498 p.

Bull, P.A. 1977. A simple peel technique for silts and clays: Jour. Sed. Petrol., v. 47, p. 1361–1362.

Bruce, C.H. 1973. Pressured shale and related sediment deformation: Mechanism for development of regional contemporaneous faults: Amer. Assoc. Petrol. Geol. Bull., v. 57, p. 878–886.

Burgess, J.D. 1974. Microscopic examination of kerogen (dispersed organic matter) in petroleum exploration: Geol. Soc. Amer. Spec. Paper 153, p. 19–29.

Burke, K. 1977. Aulacogens and continental breakup: Ann. Rev. Earth Planet. Sci., v. 5, p. 371–396.

Burst, J.F. 1965. Subaqueously formed shrinkage cracks in clay: Jour. Sed. Petrol., v. 35, p. 348–353.

———. 1976. Argillaceous sediment dewatering: Ann. Rev. Earth Planet. Sci., v. 4, p. 293–318.

Byers, C.W. 1977. Biofacies pattern in euxinic basins: A general model. Soc. Econ. Paleontol. Mineral. Spec. Publ. 25, p. 5–17.

Campbell, C.V. 1967. Lamina, laminaset, bed and bedset. Sedimentology, v. 8, p. 7–26.

Carroll, D. 1970. Clay Minerals: A Guide to Their X-Ray Identification. Geol. Soc. Amer. Spec. Paper 126, 80 pp.

Carver, R.E. Ed. 1971. Procedures in Sedimentary Petrology. Wiley–Interscience, New York, 653 pp.

Chaloner, W.G. 1968. The paleoecology of fossil spores. In: E.T. Drake, Ed., Evolution and Environment. Yale University Press, New Haven and London, p. 125–138.

Claypool, G.E., A.H. Love, and E.K. Maughan. 1978. Organic geochemistry, incipient metamorphism, and oil generation in black shale members of Phosphoria Formation, western interior United States: Amer. Assoc. Petrol. Geol. Bull., v. 62, p. 98–120.

Cloud, P. 1971. Adventures in Earth History. W.H. Freeman and Co., San Francisco, 992 pp.

Cody, R.D. 1971. Adsorption and reliability of trace elements as environment indicators of shales: Jour. Sed. Petrol., v. 41, p. 461–471.

Cole, R.D., and M.D. Picard. 1975. Primary and secondary sedimentary structures in oil shale and other fine-grained rocks, Green River Formation (Eocene), Utah and Colorado: Utah Geol., v. 2, p. 49–67.

Coleman, J.M., and L.D. Wright. 1975. Modern river deltas: Variability of processes and sandbodies. In: M.L. Broussard, Ed., Deltas: Models for Exploration. Houston Geol. Soc., Houston, Texas, p. 99–150.

Colton, G.W. 1967. Orientation of carbonate concretions in the Upper Devonian of New York. U.S. Geol. Survey Prof. Paper 575B, p. 57–59.

Compton, R.P. 1962. Manual of Field Geology. John Wiley and Sons, New York, 378 pp.

Connan, J. 1974. Time–temperature relation in oil genesis: Amer. Assoc. Petrol. Geol. Bull., v. 58, p. 2516–2521.

Conybeare, C.E.B., and K.A.W. Crook. 1968. Manual of Sedimentary Structures. Australian Dept. Nat. Develop., Bur. Mineral. Res. Geol. Geophys., v. 102, 327 pp.

Cook, H.E., P.D. Johnson, J.C. Matti, and I. Zemmels. 1975. Methods of sample preparation and X-ray diffraction data analysis, X-ray Mineralogy Laboratory, Deep-Sea Drilling Project, University of California, Riverside: In: D.E. Hayes, L.A. Frakes, P.J. Barrett, D.A. Burns, P-H. Chen, A.B. Ford, A.G. Kaneps, E.M. Kemp, D.W. McCollum, D.J.W. Piper, R.E. Wall, and P.N. Webb, Eds. Initial Reports of the Deep-Sea Drilling Project. U.S. Govt. Printing Office, Washington, D.C., v. 28, p. 999–1008.

Craig, G.Y., and E.K. Walton. 1962. Sedimentary structures and paleocurrent directions from the Silurian rocks of Kirkcudbrightshire: Edinburgh Geol. Soc. Trans., v. 19, p. 100–119.

Crimes, T.P., and J.C. Harper, Eds. 1970. Trace Fossils. Seel House Press, Liverpool, 547 pp.

————, Eds. 1977. Trace Fossils-2. Seel House Press, Liverpool, 351 pp.

Curray, J.R., and D.G. Moore. 1974. Sedimentary and tectonic processes in the Bengal deep-sea fan and geosyncline. In: C.A. Burk and C.L. Drake, Eds., The Geology of Continental Margins. Springer-Verlag, New York, p. 617–628.

Curtis, C.D. 1977. Sedimentary geochemistry: Environments and processes dominated by involvement of an aqueous phase: Trans. Roy. Soc. London, v. 286, p. 353–372.

Dimitrijević, M.N., M.D. Dimitrijević, and B. Radosĕvic. 1967. Sedimente Teksture u Turbiditima: Zavoid Geol. Geofiž. Istrazivanja, v. 16, 70 pp.

Donaldson, A.C. 1974. Pennsylvanian sedimentation of central Appalachians. In: G. Briggs, Ed., Carboniferous of the Southeastern United States. Geol. Soc. Amer. Spec. Paper 148, p. 47–78.

Dow, W.G. 1978. Petroleum source beds on continental slopes and rises: Amer. Assoc. Petrol. Geol. Bull., v. 62, p. 1584–1606.

Dżulyński, S. 1963. Directional Structures in Flysch. Polska Akad. Sci., Studia Geol. Pol., v. 12, 136 pp. (Polish and English.)

————, and J.E. Sanders. 1962. Current marks on firm mud bottoms: Trans. Connecticut Acad. Arts Sci., v. 42, p. 57–96.

————, and E.K. Walton. 1965. Sedimentary Features of Flysch and Greywackes. Developments in Sedimentology, v. 7. Elsevier Scientific Publ. Co., Amsterdam, 1965, 274 pp.

Eaton, G.P. 1978. Volcanic ash deposits. In: R.W. Fairbridge and J. Bourgeouis, Eds., The Encyclopedia of Sedimentology. Dowden, Hutchinson, and Ross, Stroudsburg, Pa., p. 843–850.

Edzwald, J.K., and C.R. O'Melia. 1975. Clay distribution in recent estuarine sediments: Clays and Clay Minerals, v. 23, p. 39–44.

Eisbacher, G.H., M.A. Carrigy, and R.B. Campbell. 1974. Paleodrainage patterns and late orogenic basins of the Canadian Cordillera. In: W.R. Dickenson, Ed., Tectonics and Sedimentation. Soc. Econ. Paleontol. Mineral. Publ. 22, p. 143–166.

Embry, A., and J.E. Klovan. 1976. The Middle-Upper Devonian clastic wedge of the Franklinian geosyncline: Bull. Canadian Petrol. Geol., v. 24, p. 485–639.

Englund, J.-O., and P. Jorgensen. 1973. A chemical classification system for argillaceous sediments and factors controlling their composition: Geol. Fören Stockholm Förh., v. 95, p. 87–97.

Epstein, A.G., J.B. Epstein, and L.D. Harris. 1977. Conodont Color Alteration—An Index to Organic Metamorphism. U. S. Geol. Survey Prof. Paper 995, 27 pp.

Eslinger, E.V., S.M. Savin, and H.-W. Yeh. 1979. Oxygen isotope geothermometry of diagenetically altered shales. In: Aspects of Diagenesis, P.A. Scholle and P.R. Schluger, Eds.: Soc. Econ. Paleont. Mineral. Sp. Pub. 26, p. 113–124.

Evans, G. 1975. Intertidal flat deposits of the North Sea. In: R.N. Ginsburg, Ed., Tidal Deposits. Springer-Verlag, New York, p. 13–20.

Fenner, P., and A.F. Hagner. 1967. Correlation of variations in trace elements and mineralogy of the Esopus Formation, Kingston, New York: Geochim. Cosmochim. Acta, v. 31, p. 237–261.

Ferm, J.C. 1970. Allegheny deltaic deposits. In: J.P. Morgan, Ed., Deltaic Sedimentation—Modern and Ancient. Soc. Econ. Paleontol. Mineral. Spec. Publ. 15, p. 246–255.

Ferguson, L. 1963. Estimation of the compaction factor of a shale from distorted brachiopod shells: Jour. Sed. Petrol., v. 33, p. 796–798.

Fisher, I.S. 1977. Distribution of Mississippian geodes and geodal minerals in Kentucky: Econ. Geol., v. 72, p. 864–869.

Fisher, W.L., and J.H. McGowen. 1967. Depositional systems in the Wilcox Group of Texas and their relationship to occurrence of oil and gas: Trans. Gulf Coast Assoc. Geol. Socs., v. 17, p. 105–125.

————, and L.F. Brown. 1972. Clastic Depositional Systems—A Genetic Approach to Facies Analysis. Texas Bur. Econ. Geol., 211 pp.

Fisk, H.N. 1947. Fine-Grained Alluvial Deposits and their Effects on Mississippi River Activity. U.S. Army Corps. Engineers, Mississippi River Commission, Vicksburg, Miss., 82 pp.

————, E. McFarlan Jr., and C.R. Kolb. 1954. Sedimentary framework of the modern Mississippi Delta. Jour. Sed. Petrol., v. 24, p. 76–99.

Flint, R.F. 1971. Glacial and Quaternary Geology. John Wiley and Sons, New York, 892 pp.

Folger, D.W. 1972. Texture and organic carbon content of bottom sediments in some estuaries of the U.S. *In:* B.W. Wilson, Ed., Environmental Framework of Coastal Plain Estuaries. Geol. Soc. Amer. Mem. 133, p. 391–408.

Folk, R.L. 1962. Petrography and origin of the Silurian Rochester and McKenzie Shales, Morgan County, West Virginia: Jour. Sed. Petrol., v. 32, p. 539–578.

———. 1968. Petrology of Sedimentary Rocks. Hemphill's Bookstore, Austin, Texas, 170 pp.

Franks, P.C. 1969. Nature, origin and significance of cone-in-cone structures in the Kowa Formation (Early Cretaceous), North-Central Kansas: Jour. Sed. Petrol., v. 39, p. 1438–1454.

Fraser, G.S., and A.T. James. 1969. Radiographic Exposure Guides for Mud, Sandstone, Limestone and Shale. Illinois Geol. Survey Circ. 443, 20 pp.

Frey, R.W. 1975. The Study of Trace Fossils. Springer-Verlag, New York, Heidelberg, Berlin, 562 pp.

Friedman, G.M. 1975. Identification of carbonate minerals by staining methods: Jour. Sed. Petrol., v. 29, p. 87–97.

Friend, P.F. 1966. Clay fractions and colours of some Devonian red beds in the Catskill Mountains, U.S.A.: Quart. Jour. Geol. Soc. London, v. 122, p. 273–292.

Fritz, P., P.L. Binda, F.E. Folinsbee, and H.R. Krause. 1969. Isotopic composition of diagenetic siderites from Cretaceous sediments in western Canada: Jour. Sed. Petrol., v. 41, p. 282–288.

Füchtbauer, H., and G. Müller. 1970. Sedimente und Sedimentgestein: Teil II Sediment-Petrologie. E. Schweizerbart'sche Verlagsbuchhandlung, Stuttgart, 729 pp.

Gall, J.C. 1976. Environments Sédimentaires Anciens et Milieux de Vie. Doin Editeurs, Paris, 228 pp.

Galloway, W.E., and L.F. Brown Jr. 1972. Depositional System and Shelf–Slope Relationships in Upper Pennsylvanian Rocks, North-Central Texas. Texas Bur. Econ. Geol., Rept. Invest. 75, 62 pp.

Gardner, W.S., and D.W. Menzel. 1974. Phenolic aldehydes as indicators of terrestrially derived organic matter in the sea: Geochim. Cosmochim. Acta, v. 38, p. 813–822.

Garrells, R.M., and C.L. Christ. 1965. Solutions, Minerals and Equilibria. Harper and Row, New York, 450 pp.

Gary, M., R. McAfee Jr., and C.L. Wolf Eds. 1972. Glossary of Geology. Amer. Geol. Inst., Washington, D.C., 805 pp.

Gibbs, R.J. 1973. The bottom sediments of the Amazon shelf and tropical Atlantic Oceans: Mar. Geol., v. 14, p. M39–M45.

———. 1977. Clay mineral segregation in the marine environment: Jour. Sed. Petrol., v. 42, p. 237–243.

Gilman, R.A., and W.J. Metzger. 1967. Cone-in-cone concretions from western New York: Jour. Sed. Petrol., v. 37, p. 87–95.

Ginsburg, R.N., Ed., 1975. Tidal Deposits. Springer Verlag, New York, 428 pp.

Glennie, K.W. 1970. Desert Sedimentary Environments. Developments in Sedimentology, v. 14, Elsevier Scientific Publ. Co., Amsterdam, 222 pp.

Goddard, E.N., P.D. Trask, R.K. Deford, O.N. Rove, J.T. Singlewald Jr., and R.M. Overbeck. 1975. Rock-Color. Chart Geol. Soc. Amer., Boulder, Colorado.

Goldhaber, M.B., and I.R. Kaplan. 1975. Controls and consequences of sulfate reduction rates in marine sediments: Soil Sci., v. 119, p. 42–55.

Grew, E.S. 1974. Carbonaceous material in some metamorphic rocks of New England and other areas: Jour. Geol., v. 82, p. 50–73.

Grim, R.E. 1968. Clay Mineralogy. 2nd ed. McGraw-Hill Book Co., New York, 596 pp.

Groat, G.C. 1972. Presidio Bolson, Trans-Pecos Texas and Adjacent Mexico. Geology of a Desert Basin Aquifer. Texas Bur. Econ. Geol., Rept. No. 76, 46 pp.

Gubler, Y., D. Bugnicourt, J. Faber, B. Kubler, and R. Nyssen. 1966. Essai de Nomenclature et Caractérisation des Principales Structures Sédimentaires. Éditions Technip, Paris, 291 pp.

Hakes, W.G. 1976. Trace Fossils and Depositional Environment of Four Clastic Units, Upper Pennsylvanian Megacyclothems, Northeast Kansas. Univ. Kansas Paleontol. Contrib. Art. 63, 46 pp.

Hallam, A. 1967. Siderite and calcite-bearing concretionary nodules in the Lias of Yorkshire: Geol. Mag., v. 104, p. 222–227.

Hampton, M.A. 1972. The role of subaqueous debris flow in generating turbidity currents: Jour. Sed. Petrol., v. 42, p. 775–793.

Hanshaw, B.B., and T.B. Coplen. 1973. Ultrafiltration by a compacted clay membrane; II—Sodium ion exclusion at various ionic strengths: Geochim. Cosmochim. Acta, v. 37, p. 2311–2327.

Haq, B.U., and Anne Boersma, Eds. 1978. Introduction to Marine Micropaleontology. Elsevier Scientific Publ. Co., New York, Oxford, 376 pp.

Harms, J.C. 1970. Why study cores? In: Mesozoic Core Seminar. Regina, Saskatchewan. Saskatchewan Geol. Soc. Dept. Mineral Res., 28–30 October, 1970, pp. 1–2.

Hattin, D.E. 1967. Stratigraphy of the Graneros Shale (Upper Cretaceous) in central Kansas. Kansas Geol. Survey Bull. 178, 83 pp.

Hayes, J.B. 1964. Geodes and concretions from the Mississippian Warsaw Formation, Keokuk Region, Iowa, Illinois, Missouri: Jour. Sed. Petrol., v. 34, p. 123–133.

Heath, G.R. 1974. Dissolved silica in deep-sea sediments: In: W.M. Hay, Ed., Studies in Paleo-Oceanography. Soc. Econ. Paleontol. Mineral. Spec. Publ. No. 20, p. 77–93.

Hedberg, H.D. 1974. Relation of methane generation to undercompacted shales, shale diapirs, and mud volcanoes: Amer. Assoc. Petrol. Geol. Bull., v. 58, p. 661–673.

Hedges, J.I., and P.L. Parker. 1976. Land-derived organic matter in surface sediments from the Gulf of Mexico: Geochim. Cosmochim. Acta, v. 40, p. 1019–1029.

Hesse, R. 1975. Turbiditic and non-turbiditic mudstone of Cretaceous flysch sections of the east Alps and other basins: Sedimentology, v. 22, p. 387–416.

———. 1977. Softex-radiographs of sliced piston cores from the Japan and southern Kurile Trench and slope areas. In: E. Honza, Ed., Geological Investigation of Japan-and Southern Kurile Trench and Slope Areas GH-76-2 Cruise, April–June 1976. Geol. Surv. Japan, Cruise Rept. No. 7 Chapter X, p. 86–108.

Hester, N.C., and Pryor, W.A. 1972. Blade-shaped crustacean burrows of Eocene age: A composite form of Ophiomorpha: Geol. Soc. Amer. Bull., v. 83, p. 677–688.

Hodgson, G.A. 1966. Carbon and oxygen isotope ratios in diagenetic carbonates from marine sediments: Geochim. Cosmochim. Acta, v. 30, p. 1223–1233.

Hoffman, P. 1973. Evolution of an Early Proterozoic continental margin: The Coronation geosyncline and associate aulacogens of the northwestern Canadian Shield: Phil. Trans. Roy. Soc. London, Ser. A., v. 273, p. 547–581.

Holeman, J.N. 1968. The sediment yield of major rivers of the world. Water Resources Research, v. 4, p. 737–747.

Hounslow, A.R. 1979. Modified gypsum/anhydrite stain. Jour. Sediment. Petrology, v. 49, p. 636–637.

Howard, J.D., Frey, R.W., and Reineck, H.-E. 1972. Georgia coastal region, Sapelo Island, U.S.A.: Sedimentology and biology: Senckenbergiana Maritima, v. 4, 223 pp.

Hudson, J.D. 1977. Stable isotopes and limestone lithification: Jour. Geol. Soc. London, v. 133, p. 637–660.

———. 1978. Concretions, isotopes, and the diagenetic history of the Oxford clay (Jurassic) of central England: Sedimentology, v. 25, p. 339–370.

Hutchinson, G.E. 1957. A Treatise on Limnology, v. 2, Introduction to Lake Biology and the Limnoplankton. John Wiley and Sons, New York, 1115 pp.

Imbrie, J., and N.D. Newell Eds. 1964. Approaches to Paleoecology. John Wiley and Sons, New York, 432 pp.

Ingram, R.L. 1954. Terminology for thickness of stratification and parting units in sedimentary rocks: Geol. Soc. Amer. Bull., v. 65, 1954, pp. 937–938.

Jacob, A.J. 1973. Elongate concretions as paleochannel indicators, Tongu River Formation (Paleocene) North Dakota: Geol. Soc. Amer. Bull., v. 84, p. 2127–2132.

Jaworowski, K. 1971. Sedimentary structures of the Upper Silurian siltstones in the Polish lowland: Acta Geol. Pol., v. 21, p. 519–571.

Jones, D.J. 1956. Introduction to Microfossils. Harper and Bros., New York, 406 pp.

Jones, M.L., and J.A. Clendening. 1968. A feasibility study for paleocurrent analysis in lutaceous Monongahela–Dunkard strata of the Appalachian Basin: Proc. West Virginia Acad. Sci., v. 40, p. 225–261.

———, and J.M. Dennison. 1970. Oriented fossils as paleocurrent indicators in Paleozoic lutites of southern Appalachians: Jour. Sed. Petrol., v. 40, p. 642–649.

Jüngst, H. 1934. Geologic significance of synaeresis: Geol. Rundschau, v. 23, p. 321–325.

Kelly, K.L., and D.B. Judd. 1976. Color, Universal Language and Dictionary of Names. U.S. Dept. Commerce, Natl. Bur. Standards, Spec. Publ. 440, 158 pp.

Kemper, E., and W. Zimmerle. 1978. Die anoxischen Sedimente der präoberaptischen Unterkreide NW-Deutschlands und ihr paläogeographischen Rahmen. Geol. Jb., Reihe A., v. 45, p. 3–41.

Khabakov, A.V. Ed. 1962. Atlas Tekstur i Struktur Osadochnykh Gornykh Porod. (An Atlas of Textures and Structures of Sedimentary Rocks. Part. 1, Clastic and Argillaceous Rocks.) Vsegei, Moscow, 578 pp. (Russian with French translation of plate captions.)

Kieken, M. 1973. Evolution de l´Aquitaine au cours du Tertiare: Soc. Géol. France, Ser. 7, v. 15, pp. 40–50.

Klein, G. deV. 1967. Comparison of recent and ancient tidal flat and estuarine sediments: In: G.H. Lauff, Ed., Estuaries. Horn-Schafer, Baltimore, p. 207–218.

Klemme, H. D. 1971. Giants, supergiants and their relation to basin types: Oil Gas Jour., 1, 8, and 15 March.

logy, v. 23, Elsevier Scientific Publ. Co., Amsterdam, p. 57–76.

1. Examination of well cuttings: Quart. Colorado School Mines, v. 46,

Examination of well cuttings and the lithologic log. *In:* L.W. LeRoy
LeRoy, Eds., Surbsurface Geology. Colorado School of Mines, Golden,
p. 286–304.

.D., and Samuels, N.D. 1980. Field classification of fine-grained
ur. Sed. Petrol., v. 50.

1974. Significance of color in red, green, purple, olive, brown, and gray
DiFunta Group, Northeastern Mexico: Jour. Sed. Petrol., v. 44, p.

, and L.S. Yeakel. 1963. Relationship between parting lineation and
ic: Jour. Sed. Petrol., v. 33, p. 779–782.

1971. Wave effectiveness at the sea bed and its relationship to bed-
d deposition of mud: Jour. Sed. Petrol., v. 41, p. 89–96.

2. Transport and escape of fine-grained sediment from shelf areas. *In:*
witt, Ed. Shelf Sediment Transport: Process and Pattern. Dowden,
son and Ross, Inc., Stroudsburg, Pa., p. 225–248.

5. Vertical flux of particles in the ocean: Deep-Sea Res., v. 22, p.
2.

W.G., B.J. Goldman, and P. Payne., Eds. 1968. Deserts of the World.
rizona Press, Tuscon 788 pp.

, and G.W. Weir. 1953. Terminology for stratification and cross-
ation in sedimentary rocks: Geol. Soc. Amer. Bull., v. 64, p. 381–390.

F.T. 1975. Sedimentary cycling and the evolution of sea water. *In:* J.P.
d G. Skirrow, Eds. Chemical Oceanography, 2nd ed., Academic Press,
, New York, p. 309–364.

W.S. 1972. Fibrous calcite, a Middle Devonian geologic marker, with
aphic significance, District of MacKenzie, Northwest Territories: Canadi-
. Earth Sci., v. 9, p. 1431–1440.

1975. Reevaluation of montmorillonite dehydration as cause of abnormal
e and hydrocarbon migration: Amer. Assoc. Petrol. Geol. Bull., v. 59, p.
02.

1959. The Composite Interpretive Method of Logging Drill Cuttings.
oma Geol. Survey Guide 8, 48 pp.

P.E. Litz, and W.D. Huff. 1979. A new technique for making thin sections
ey sediments. Jour. Sed. Petrol., v. 49, p. 641–643.

, A. 1965. Aspects of a Middle Cambrian thanatope of Oland: Geol.
Stockholm Förh., v. 87, p. 181–230.

70. Toponomy of trace fossils. *In:* T.P. Crimes and J.C. Harper, Eds. Trace
s. Geol. Jour., Spec. Issue 3, p. 323–330.

V., and Howard J.D., Eds., 1975. Estuaries of the Georgia coast, U.S.A.
lentology and biology, VI. Animal–sediment relationships of a salt-marsh
y—Doboy Sound: Senckenbergiana Maritima, v. 7, p. 205–206.

, G.V., and M.A. Hampton. 1973. Sediment gravity flows: Mechanics of
nd deposition. *In:* Turbidites and Deep Water Sedimentation. Soc. Econ.
ntol. Mineral. Pacific Section Short Course, Anaheim, Calif., 157 pp.

. 1968. Étude des proprietés physiques de différents sédiments tres fin et
omportement sous des actions hydrodynamiques: La Houille Blanche,

Krauskopf, K.B. 1979. Introduction t(...
New York, N.Y., 617 pp.

Krinitzsky, E.L. 1970. Correlation of Ba...
VI, Atchafalaya Levee System, Lo...
terways Exp. Sta. Tech. Rept. S-7(...
———, and F.L. Smith. 1969. Geology...
Basin, Louisiana. U.S. Army Corp...
Rept. S-69-8, 58 pp.

Krone, R.B. 1962. Flume Studies of the...
Processes. Hydraulic Eng. Lab. an(...
at Berkeley, 110 p.

Krumbein, W.C. 1960. The "geological...
numerical data in geology: Geol....
341–368.

Krynine, P.D. 1948. The megascopic st(...
rocks: Jour. Geol., v. 56, p. 130–16...

Kübler, B. 1966. La cristallinité de l'illite (...
morphisme: In: Colloque sur les Éta...
nière, Neuchâtel, p. 105–122.

Kuenen, Ph. H. 1958. Experiments in geo...
1–28.

Kummel, B., and B. Raup, Eds. 1965. H...
W.H. Freeman and Co., San Francisc...

LaBarbera, M. 1977. Brachiopod orientati(...
tory behavior and field orientations: I...

Lanteume, M., B. Beaudoin, and R. Cam...
flysch. Grés d´Annot: du synclinal de I...
Recherch Scientifique, Paris, 97 pp.

Larese, R.E., and Heald, M.T. 1977. Petro(...
Samples from the CGTC 20403 and C(...
Counties, West Virginia. U.S. Dept. of...

Lerman, A. 1975. Maintenance of steady stat...
v. 275, p. 609–635.
———, 1979. Geochemical Processes: W...
Wiley and Sons, New York, 481 pp.

LeRoy, D. 1977. Economic microbiostratigr(...
Eds., Subsurface Geology. Colorado !...
p. 212–233.

Leventhal, J., S.E. Suess, and P. Cloud. 1975...
in kerogen from Pre-Phanerozoic sedim...
ton, D.C., v. 72, p. 4706–4710.

Lewan, M.D. 1978. Laboratory classification (...
Geology, v. 8, p. 745–748.

Lindström, M. 1964. Conodonts. Elsevier Scie...
——— and W. Ziegler, Eds. 1972. Conodont...
pp.

Londsdale, P., and F.N. Spiess. 1977. Abyss...
towed instrument package. In: B.C. Heez(...
tion on Sedimentary Accumulations in...

v. 7, p. 59–62.

———. 1977. Action des courants, de la houille du vent sur les sediments: La Houille Blanche, v. 18, p. 9–47.

Millot, G. 1978. Clay genesis. *In:* R.H. Fairbridge and J. Bourgeois, Eds., The Encyclopedia of Sedimentology, v. VI. Dowden, Hutchinson and Ross, Inc., Stroudsburg, Pa. p. 591–620.

Mitchum, R.M. Jr., P.R. Vail, and J.B. Sangree. 1977. Seismic stratigraphy and global changes of sea level, Part 6: Stratigraphic interpretation of seismic reflection patterns in depositional sequences. *In:* C.E. Payton, Ed., Seismic Stratigraphy—Applications to Hydrocarbon Exploration. Amer. Assoc. Petrol. Geol. Mem. 26, p. 117–133.

Moon, C.F. 1972. The microstructure of clay sediments: Earth Sci. Rev., v. 8, n. 3, p. 303–321.

Moore, C.A. 1963. Handbook of Subsurface Geology. Harper and Row, New York, Evanston, and London, 235 pp.

Moore, H.B., and P.A. Kruse. 1956. A Review of Present Knowledge of Faecal Pellets. Miami Univ. Inst. Marine Sci. Marine Lab. Rept. 13806, 25 pp.

Morris, S. Conway. 1977. Fossil Priapulid Worms: Paleon. Assoc. Sp. Papers in Paleontology, n. 20, 95 pp.

Morris, S. Conway, and H.B. Whittington. 1979. The animals of the Burgess Shale: Scientific American, v. 241, p. 122–135.

Moors, H.T. 1959. The position of graptolites in turbidites: Sed. Geol., v. 3, p. 241–261.

Munsell Color Co. 1975. Munsell Soil Color Charts, for Use of Soil Scientists, Geologists, and Archaeologists. Munsell Color, Baltimore, Md.

Murata, K.J., I. Friedman, and J.D. Gleason. 1977. Oxygen isotope relations between diagenetic silica minerals in Monterey Shale, Tembler Range, California: Amer. Jour Sci., v. 277, p. 259–272.

Murray, H.H. 1954. Genesis of clay minerals in some Pennsylvanian shales of Indiana and Illinois: Clays Clay Minerals, p. 49–67.

Nanz, R.H. 1953. Chemical composition of PreCambrian slates with notes on the geochemical evolution of lutites: Jour. Geol., v. 61, p. 51–64.

Natland, M.L. 1933. Depth and temperature distribution of some recent and fossil Foraminifera in the Southern California region: Bull Scripps Inst. Oceanogr., La Jolla, Tech. Ser. 3, p. 225–230.

Neale, J.W., Ed. 1967. The Taxonomy, Morphology, and Ecology of Recent Ostracoda. Oliver and Boyd, Ltd., Edinburgh, 553 pp.

Newman, J.W., P.L. Parker, and E.W. Behrens. 1973. Organic carbon isotope ratios in Quarternary cores from the Gulf of Mexico: Geochim. Cosmochim. Acta, v. 37, p. 225–238.

Nixon, R.P. 1973. Oil source beds in the Cretaceous Mowry Shale of northwestern interior United States: Amer. Assoc. Petrol. Geol. Bull., v. 57, p. 136–161.

O'Brien, D.E., and M.E. Chenevert. 1973. Stabilizing sensitive shales with inhibited potassium-based drilling fluids: Trans. Soc. Petrol. Eng. (SPE-AIME), v. 255, p. 1089–1100.

Odell, J. 1977. LOGGER, a package which assists in the construction and rapid display of stratigraphic columns from field data: Comput. Geosci., v. 3, p. 349–379.

Odum, I.E. 1967. Clay fabric and its relation to structural properties in mid-continent

Amer. Geol., v. 19, p. 367–391.

Rusnak, G.A. 1960. Sediments of Laguna Madre, Texas. *In:* F.P. Shepard, F.B. Phleger, and Tj. H. van Andel, Eds., Recent Sediments, Amer. Assoc. Petrol. Geol., p. 153–196.

Russell, K.L. 1970. Geochemistry and halmyrolysis of clay minerals, Rio Ameca, Mexico: Geochim. Cosmochim. Acta, v. 34, p. 893–907.

Ryan, W.B.F., and M.B. Cita. 1977. Ignorance concerning episodes of ocean-wide stagnation. *In:* B.C. Heezen, Ed., Influence of Abyssal Circulation on Sedimentary Accumulations in Space and Time. Developments in Sedimentology, v. 23. Elsevier Scientific Publ. Co., Amsterdam, p. 197–215.

Schäfer, W. 1972. Ecology and Palaeoecology of Marine Environments. Oliver and Boyd, Edinburgh, 568 pp. (Translated from the German by I. Dertel.)

Schmidt, G.W. 1973. Interstitial water composition and geochemistry of deep Gulf Coast shales and sandstones: Amer. Assoc. Petrol. Geol. Bull., v. 57, p. 321–331.

Schneider, E.D. 1972. Sedimentary evolution of rifted continental margins. *In:* R. Shagam, R.B. Hargraves, W.J. Morgan, F.B. Van Houten, C.A. Burk, H.D. Holland, and L.C. Hollister, Eds. Studies in Space and Planetary Sciences. Geol. Soc. Amer. Mem. 132, p. 109–118.

Schopf, J.W., K.A. Kvenvolden, and E.S. Barghoorn. 1968. Amino acids in Precambrian sediments: An assay: Proc. Natl. Acad. Sci., Washington, D.C., v. 59, p. 639–646.

Scott, A.J., and W.L. Fisher. 1969. Delta systems and deltaic deposition. *In:* W.L. Fisher, Ed., Delta Systems in Exploration for Oil and Gas. A Research Colloquium, August 27-29, 1969. Texas Bur. Econ. Geol., p. 10–39.

Scruton, P.D. 1960. Delta building and the deltaic sequence. *In:* F.P. Shepard, F.B. Phleger, and Tj. H. van Andel, Eds., Recent Sediments, Northwest Gulf of Mexico. Amer. Assoc. Petrol. Geol., p. 82–102.

Sedimentation Seminar. 1978. Studies for students: A question set for sands and sandstones: Brigham Young Univ. Geol. Studies, v. 24, p. 1–8.

Seilacher, A. 1953. Studien zur Palichnologie. I. Über die Methoden der Palichnologie. Neues Jahrb. Geol. Palaont., Abh. 98, p. 87–124.

———. 1960. Strömungsanzeichen im Hunsrückschiefer: Notizbl. Hess. Landesamt Boden forsch. Weisbaden, v. 88, p. 88–106.

———. 1973. Biostratinomy: The sedimentology of biologically standardized particles. *In:* R.N. Ginsburg, Ed., Evolving Concepts in Sedimentology. John Hopkins Univ. Press, Baltimore and London, p. 159–177.

———, und D. Meischner. 1964. Fazies-analyse im Paläozoikum des Oslo-Gebietes: Geol. Rundschau, v. 54, p. 596–619.

Selley, R.C. 1976. An introduction to Sedimentology. Academic Press, New York, 408 pp.

Sengör, A.M.C., K. Burke, and J.F. Dewey. 1978. Rifts at high angles to orogenic belts: Tests for their origin and the Upper Rhine Graben as an example: Amer. Jour. Sci., v. 278, p. 24–40.

Scott, A.J., and W.L. Fisher. 1969. Delta systems and deltaic deposition, discussion notes. *In:* W.L. Fisher, L.F. Brown, Jr., A.J. Scott, and J.H. McGowen, Eds. Delta Systems in the Exploration of Oil and Gas, A Research Colloquium, 27-29, August, 1969. Texas Bur. Econ. Geol., p. 10–39.

Shaw, D.M. 1956. Geochemistry of pelitic rocks. III: Major elements and general geochemistry: Geol. Soc. Amer. Bull., v. 67, p. 919–934.

Shaw, D.B., and C.E. Weaver. 1965. The mineralogical composition of shales: Jour. Sed. Petrol., v. 35, p. 213–222.

Shelton, J.W. 1962. Shale compaction in a section of Cretaceous Dakota Sandstone, northwestern North Dakota: Jour. Sed. Petrol., v. 32, p. 874–877.

Sherman, G.D. 1952. The genesis and morphology of alumina-rich laterite clays. *In:* Problems of Clay and Laterite Genesis. Amer. Inst. Min. Metal., New York, p. 154–161.

Shimp, N.F., J. Witters, P.E. Potter, and J.A. Schleicher. 1969. Distinguishing marine and freshwater Muds: Jour. Geol., v. 77, p. 566–580.

Shultz, D.J., and J.A. Calder. 1976. Organic carbon $^{13}C/^{12}C$ variations in estuarine sediments: Geochim. Cosmochim. Acta, v. 40, p. 381–385.

Simonson, R.W., and C.E. Hutton. 1954. Distribution curves for loess. Amer. Jour. Sci., v. 252, p. 99–105.

Simpson, S. 1975. Classification of trace fossils. *In:* R.W. Frey, Ed., The Study of Trace Fossils. Springer-Verlag, New York, Berlin, p. 39–54.

Société Nationale Elf-Aquitaine (Production). 1977. Bulletin Centres Recherches Exploration-Production Elf Aquitaine, v. 1, p. 109–320.

Southard, J.B. 1974. Erodibility of fine abyssal sediment. *In:* A.L. Inderbitzen, Ed., Deep-Sea Sediments. Plenum Press, New York and London, p. 367–379.

Spears, D.A. 1973. Relationship between exchangeable cations and paleosalinity: Geochim. Cosmochim. Acta, v. 37, p. 77–85.

Spencer, A.M. Ed. 1974. Mesozoic-Cenozoic Orogenic Belts. Scottish Academic Press and Geol. Soc. Spec. Publ. 4, Edinburgh and London, 809 pp.

Stahl, W. 1975. Kohlenstoff-isotopenverhältnisse von Erdgassen: Riefekennzeichen ihrer Muttersubstanzen: Erdöl und Kohle, v. 28, p. 188–191.

Stainforth, R.M., J.L. Lamb, H. Luterbacher, J.H. Beard, and R.M. Jeffords. 1975. Cenozoic Planktonic Foraminiferal Zonations and Characteristics of Index Forms. Univ. Kansas Paleon, Contrib. Art. 62, 425 pp.

Staplin, F.L. 1969. Sedimentary organic matter, organic metamorphism, and oil and gas occurrence: Canadian Petrol. Geol. Bull., v. 17, p. 47–66.

Stonecipher, C.A. 1976. Origin, distribution and diagenesis of phillipsite and clinoptilolite in deep-sea sediments: Chem. Geol., v. 17, p. 307–318.

Sundborg, Å. 1956. The river Karälven: A study of fluvial process: Geograf. Annal., v. 38, p. 127–316.

Sutton, R.G., Z.P. Bowen, and A.L. McAlester. 1970. Marine shelf environments of the Upper Devonian Sonyea Group of New York: Geol. Soc. Amer. Bull., v. 81, p. 2975–2992.

Swann, D.H., J.A. Lineback, and E. Frund. 1965. The Borden siltstone (Mississippian) delta in southwestern Illinois: Illinois Geol. Survey Circular 386, 20 pp.

Swift, D.J.P. 1969. Inner shelf sedimentation: Processes and products. *In:* D.J. Stanley, Ed., The New Concepts of Continental Margin Sedimentation. Amer. Geol. Inst. Short Court Notes, Lecture 4, p. DS-4-1–DS-4-4.

Tan, F.C., and J.D. Hudson. 1974. Isotopic studies on the paleoecology and diagenesis of the Great Estuarine Series (Jurassic) of Scotland: Scot. Jour. Geol., v. 10, p. 91–128.

Thayer, C.W. 1974. Marine paleoecology in the Upper Devonian of New York: Lethaia, v. 7, p. 121–155.

———. 1975. Morphologic adaptations of benthic invertebrates to soft substrata:

Jour. Mar. Res., v. 33, 177–189.

Theron, J.N. 1971. A stratigraphic study of the Bokkeveld Group (Series). Intl. Union Geol. Sciences Commission on Stratigraphy, Subcomm. Gondwana Stratigraphy, Palaeontology, Gondwana Symposium Proc. Paper No. 2, p. 197–204.

Thomas, J, and H.H. Damberger, 1976. Internal Surface Area, Moisture Content, and Porosity of Illinois Coals: Variations with Coal Rank. Illinois Geol. Survey Circular 493, 38 pp.

Thomas, W.A. 1977. Evolution of Appalachian–Ouachita salients and recesses from reentrants and promontories in the continental margin: Amer. Jour. Sci., v. 277, p. 1233–1278.

Till, R., and D.A. Spears. 1969. The determination of quartz in sedimentary rocks using an X-ray diffraction method: Clays Clay Minerals, v. 17, p. 323–327.

Tissot, B., and D.H. Welte. 1978. Petroleum Formation and Occurrence. Springer-Verlag, Berlin, Heidelberg, New York, 538 pp.

———, B. Durand, J. Espitalie, and A. Combaz. 1974. Influence of nature and diagenesis of organic matter in formation of petroleum: Amer. Assoc. Petrol. Geol. Bull., v. 58, p. 499–506.

Todd, J.E. 1903. Concretions and their geologic effects: Geol. Soc. Amer. Bull., v. 14, p. 353–368.

Tomlinson, C.W. 1916. The origin of red beds: Amer. Jour. Sci., v. 24, p. 153–179.

Toth, D.J., and A. Lerman. 1977. Organic matter reactivity and sedimentation rates in the ocean. Amer. Jour. Sci., v. 277, p. 465–485.

Tourtelot, H.A. 1960. Origin and use of the word "shale": Amer. Jour. Sci., Bradley Vol., v. 258-A, p. 335–343.

Trask, P.D., 1937. Studies of source beds in Oklahoma and Kansas: Amer. Assoc. Petrol. Geol. Bull., v. 21, p. 1377–1393.

———, and C.C. Wu. 1930. Does petroleum form in sediments at time of deposition? Amer. Assoc. Petrol. Geol. Bull., v. 14, p. 1451–1463.

Treibs, A. 1934. Chlorophyll und Hämin-derivate in bituminösen Gesteinen, Erdölen, Erdwachsen und Asphalten. Ein Beitrag zur des Erdöls: Annalen der Chemie, v. 510, p. 42–62.

Twenhofel, W.H. 1937. Terminology of the fine grained mechanical sediments. Rept. Comm. on Sedimentation for 1936–37, v. 5. Natl. Res. Council Washington, D.C., p. 81–104.

Usdowski, H.-E. 1963. Die Genese der Tutenmergel oder Nagelkalke (cone-in-cone); Beiträge Mineral. Petrograf., v. 9, p. 95–110.

Van Houten, F.B. 1969. Late Triassic Newark Group, north central New Jersey and adjacent Pennsylvania and New York, Field Trip No. 4. In: S. Subitzky, Ed., Geology of Selected Areas in New Jersey and Eastern Pennsylvania and Guidebook of Excursions. Rutgers Univ. Press, New Brunswick, N.J., p. 314–347.

———. 1973. Origin of red beds: A review—1961–1972: Ann. Rev. Earth Planet. Sci., v. 1, p. 39–42.

Van Moort, J.C. 1972. The K_2O, CaO, MgO and CO_2 content of shales and related rocks and their implications for sedimentary evolution since the Proterozoic. In: R.W. Boyle and D.M. Shaw, Conveners. Geochemistry. 24th Intl. Geol. Congr. Sec. 10, p. 427–439.

Van Olphen, H. 1978. Clay Colloid Chemistry. 2nd ed. Interscience, New York, 300 pp.

Vanoni, V.A. 1941. Some experiments on the transportation of suspended load: Amer. Geophys. Union Trans., v. 22, p. 608–621.

Van Straaten, L.M.J.U. 1961. Sedimentation in tidal flat areas: Jour. Alberta Soc. Petrol. Geol., v. 9, n. 7, p. 203–226.

Various Authors. 1974. Dresser Atlas. Dresser Atlas Division, Houston, Tx. Dresser Industries, Inc., various paging.

Vinogradov, A.P., and A.B. Ronov. 1956. Evolution of chemical composition of clays of the Russian platform: Geochemistry, v. 2, p. 123–139.

Voight, E. 1968. Über Hiatus-konkretionen (dargestellt an Beispielen aus dem Lias): Geol. Rundshau, v. 58, p. 281–296.

Waggoner, P.E., and C. Bingham. 1961. Depth of loess and distance from source: Soil Sci., v. 92b, p. 366–401.

Walker, R.G. 1967. Turbidite sedimentary structures and their relationship to proximal and distal depositional environments: Jour. Sed. Petrol., v. 37, p. 25–43.

Walker, T.R., 1967. Formation of red beds in modern and ancient deserts. Geol. Soc. Amer. Bull., v. 78, p. 353–368.

———. 1978. Deep-water sandstone facies and ancient submarine fans: Models for exploration for stratigraphic traps: Amer. Assoc. Petrol. Geol. Bull., v. 62, p. 932–966.

Wanless, H.R., J.R. Baroffio, J.C. Gamble, J.C. Horne, D.R. Orlopp, A. Rocha-Campos, J.E. Souter, P.C. Trescott, R.S. Vail, and C.R. Wright. 1970. Late Paleozoic deltas in the central and eastern United States. In: J.P. Morgan, Ed. Deltaic Sedimentation-Modern and Ancient. Soc. Econ. Paleontol. Mineral. Spec. Publ. 15, p. 215–245.

Way, D.S. 1973: Terrain Analysis. Dowden, Hutchinson and Ross, Inc. Stroudsburg, Pa., 392 pp.

Weaver, C.E. 1967. Potassium, illite, and the ocean: Geochim. Cosmochim. Acta, v. 31, p. 2181–2196.

———. 1978. Clay sedimentation facies. In: R.W. Fairbridge and J. Bourgeois, Eds., The Encyclopedia of Sedimentology, v. VI. Dowden, Hutchinson and Ross, Stroudsburg, Pa., p. 159–164.

Weber, K.J. 1971. Sedimentological aspects of oil fields in the Niger delta: Geol. Mijnbouw, v. 50, p. 559–576.

Weeks, L.G. 1953. Environment and mode of origin and facies relationship of carbonate concretions in shales: Jour. Sed. Petrol., v. 23, p. 162–173.

———. 1957. Origin of carbonate concretions in shales, Magdalena Valley, Colombia: Geol. Soc. Amer. Bull., v. 68, p. 95–102.

Wenthworth, C.K. 1922. A scale of grade and class terms for clastic sediments: Jour. Geol., v. 30, p. 377–392.

White, W.A. 1961. Colloid phenomena in sedimentation of argillaceous rocks: Jour. Sed. Petrol., v. 31, p. 560–570.

Whitten, E.H.T. 1966. Structural Geology of Folded Rocks. Rand McNally and Co., Chicago, 678 pp.

Wickwire, G.T. 1936. Crinoid stems on fossil wood: Amer. Jour. Sci., Ser. 5, v. 32, p. 145–146.

Williams, G.E. 1973. Late Quaternary piedmont sedimentation, soil formation and paleoclimates in arid southern Australia: Z. Geomorphol., v. 17, p. 102–125.

Wilson, R. L. 1975. Carbonate Facies in Geologic History. Springer-Verlag, New York, 471 pp.

Wilson, J.J., and D.R. Clark, 1978. X-ray identification of clay minerals in thin sections. Jour. Sediment. Petrology, v. 48, p. 650–660.

Windom, H.L. 1975. Eolian contribution to marine sediments: Jour. Sed. Petrol., v. 45, p. 520–529.

Woodbury, H.O., I.B. Murray Jr., P.J. Pickford, and W.H. Akers. 1973. Pliocene and Pleistocene depocenters, outer continental shelf, Louisiana and Texas: Amer. Assoc. Petrol. Geol. Bull., v. 57, p. 2428–2439.

Wray, J.L. 1977. Calcareous Algae. In: Developments in Paleontology and Stratigraphy, v. 4. Elsevier Scientific Publ. Co., Amsterdam, 185 pp.

Wright, L.D., and J.M. Coleman. 1973. Variations in morphology of major river deltas as functions of ocean wave and river discharge regimes: Amer. Assoc. Petrol. Geol. Bull., v. 57, p. 370–398.

Young, R.N., and J.B. Southard. 1978. Erosion of fine-grained marine sediments: Seafloor and laboratory experiments: Geol. Soc. Amer. Bull., v. 89, p. 663–672.

Zangerl, R., and E.S. Richardson. 1963. The Paleoecological History of Two Pennsylvanian Black Shales: Fieldiana Geol. Mem., v. 4, 352 pp.

Ziegler, A.M., and W.S. McKerrow. 1975. Silurian marine red beds: Amer. Jour. Sci., v. 275, p. 31–56.

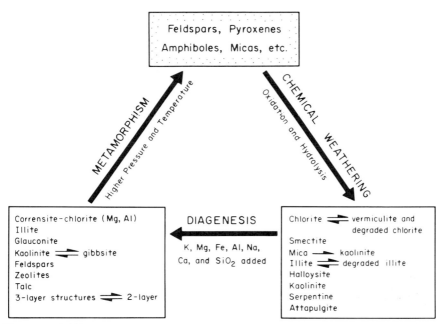

Figure 3.1 Mineralogic transformations and processes that generate the principal minerals of shales. (Redrawn from Keller 1964, Fig. 1.)

Section 4, "Silts and Clays" (pp. 131–274), is very comprehensive and includes classification; mineralogy; environments of deposition; descriptions of some modern muddy environments; much on diagenesis; a discussion of special shales, such as those that are bituminous, feldspar-rich, and aluminum rich; and a discussion of the transition of muds and shales to metamorphic rocks. See author's Fig. 4-68 (not shown), which summarizes diagenetic changes in response to burial, pressure, temperature, and duration of burial.

Grim, R.E. 1968. Clay Mineralogy. 2nd ed. McGraw-Hill Book Co., New York, 596 pp.

The standard work used by geologists and many others. Fourteen chapters, the last discussing origin and occurrence of the clay minerals.

Keller, W.D. 1964. Processes of origin and alteration of clay minerals. *In:* C.I. Rich and G.W. Kunze, Eds. Soil Clay Mineralogy; A Symposium, Univ. North Carolina Press, Chapel Hill, p. 3–76.

Informative summary of the origins of clay minerals, their alterations, diagenesis, and geologic history, by a famous clay mineralogist. Especially interesting section on soil processes in the formation of clay minerals. Extensive bibliography. (Fig. 3.1).

Millot, G. 1970. Geology of Clays. Springer-Verlag, New York, Heidelberg, Berlin, 429 pp.

Translated from the original French edition of 1964, this book sums up a lifetime of study of shales and clays in 12 chapters. Strong emphasis upon depositional environments plus a rare treatment of specific case histories. Compare with Grim (1968).

Pettijohn, F.J. 1975. Sedimentary Rocks. 3rd ed. Harper and Row, New York, 628 pp.

Chapter 8, "Shales, Argillites and Siltstones," contains about 200 references. Emphasis is on bulk chemistry and its significance. Comments on "the shale problem" are instructive. Well written.

Strakhov, N.M. 1969. Principles of Lithogenesis, v. 2. Consultants Bureau, New York and Edinburgh, 609 pp.

Excellent introduction to the Russian literature; strong emphasis on effects of climate. Pages 338–354 discuss oil shales; concretions are given an entire chapter.

van Olphen, H. 1963. Introduction to Clay Colloid Chemistry. Wiley Interscience, New York, 300 pp.

The best treatment of the behavior of clays in suspension, an essential element in understanding the sedimentation of fine-grained particles. Has a very good discussion of clay–organic interactions. Useful chapter summaries.

von Engelhardt, W.F. 1973. Die Bildung von Sedimenten und Sediment-gesteinen. Sediment-Petrologie, Teil III, Schweizerbart'sche Verlagsbuchhandlung, Stuttgart, 378 pp.

Good general treatment of deposition, compaction, and chemical diagenesis of clays. Detailed discussion of carbonate concretions in shales.

Yariv, S., and H.Cross. 1979. Geochemistry of Colloid Systems. Springer-Verlag, Berlin-Heidelberg, New York, 450 pp.

Up-to-date review of the physical chemistry of geologically important colloids. These include clays, silica, and iron-manganese hydroxides. A chapter on the colloid geochemistry of argillaceous sediments, which includes a discussion of the nature of pores in shales, is especially valuable.

CLASSICS

The papers of this section provide a glimpse of some outstanding and unusual early studies of shales by geologists. Although there are doubtlessly more, we have been hard put to find them.

Archanquelsky, A.D. 1927, 1928. Ob Osadkak Chernov Morya i ik Znachenii v Poznanii Osadochnik Gornik Porod. (On the Black Sea Sediments and Their Importance for the Study of Sedimentary Rocks.): Bull. Soc. Naturalistes de Moscou Sec. Géol., n.s., v. 5, p. 261–289, and v. 6, p. 90–108.

Description in Russian (with a long English summary) of the argillaceous sediments of the Black Sea and their general significance for the argillaceous fill of

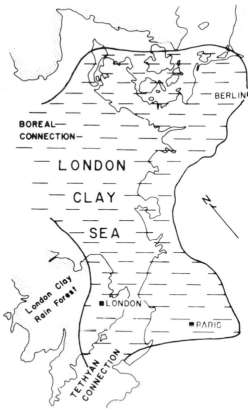

Figure 3.2 Paleogeography of the London Clay Sea—an early example of a regional analysis of a shaly basin. (Redrawn from Davis and Elliott 1957, Fig. 2.)

geosynclines by a very famous Russian sedimentologist. Volume 6 contains cross sections and maps. A truly great classic, still poorly known in much of the West.

Davis, A.G., and G.F. Elliott. 1957. The paleogeography of the London Clay Sea: Geol. Assoc. London Proc., v. 68, p. 255–277.

A very classical paper that is useful to show how basin analysis was done in years gone by. Very stratigraphic with emphasis upon paleontology, but the article's regional paleogeographic map, (Fig. 3.2) is outstanding—there are still only a few such maps of ancient muddy seas. (See also Figs. 3.55 and 3.73).

Fisher, D.W. 1957. Lithology, Paleoecology, and Paleontology of the Vernon Shale (Late Silurian) in the Type Area: New York State Mus. Sci. Serv., Bull. 364, 31 pp.

Early, integrated analysis of a shale stresses paleoecology and paleontology and relates both to five types of shale.

Grim, R.E., W.F. Bradley, and W.A. White. 1957. Petrology of the Paleozoic Shales of Illinois. Illinois Geol. Survey Rept. Invest 203, 35 pp.

Early, major clay mineralogy study of Paleozoic shales in the Illinois Basin, supplemented by large numbers of chemical analyses, optical data, electron micrographs, and thin-section studies. Although the report is now technically obsolete, the concept of a systematic petrologic inventory of shales in a given area or basin is still very, very relevant.

Keller, W.D., and C.P. Ting. 1950. The petrology of a specimen of the Perry Farm Shale: Jour. Sed. Petrol., v. 20, p. 123–132.

Early complete mineralogic description of a marine shale by famous clay mineralogists.

Millot, G. 1949. Relations entre la Constitution et la Genèse des Roches Sédimentaires Argileuses: Geol. Appl. Prospect. Minière, v. 2, 352 pp.

A classic paper organized into "Definitions and Methods," "Facts" (a discussion of different sedimentary units mostly in France and nearby Germany) and "Interpretations," wherein a strong argument is made for environmental control of clay mineralogy. Three hundred and ninety-four references, and seven plates, some showing powder photographs and microphotographs of shales in thin section. See also Millot (1970).

Moore, H.B. 1931. The muds of the Clyde Sea area. III: Chemical and physical conditions; rate and nature of sedimentation; and fauna: J. Mar. Biol. Assoc. U. K. n.s., v. 17, p. 325-358.

Pioneering study of pH, water content, density, and fecal pellets, plus sedimentation rate and fauna density versus depth. Notes that up to 40 percent of the mud can consist of fecal pellets.

Murray, J., and A.F. Renard. 1891. Report on Deep-Sea Deposits, Based on the Specimens Collected During the Voyage of H.M.S. *Challenger* in the Years 1872 to 1876. Her Majesty's Stationery Office, London, 525 pp.

A classic scientific work of art in six chapters, including the history of oceanography, methods of data collection, and nature, composition, and distribution of deep-sea sediments. The latter is most comprehensive, including organic, terrestrial, and extraterrestrial materials, plus a world map showing their distribution and 29 beautiful plates, plus 36 woodcuts and many charts. Very careful petrographic and chemical study of the dredged samples (more than 12 000), of which many were terrestrial muds. But one of 50 volumes resulting from the *Challenger's* world-wide cruise, one of the great scientific expeditions of all time.

Pepper, J.F., W. de Witt Jr., and D.F. Demarest. 1954. Geology of the Bedford Shale and Berea Sandstone in the Appalachian Basin. U.S. Geol. Survey Prof. Paper 259, 111 pp.

An interesting example of a combined subsurface and outcrop study of a shale unit, emphasizing its relationship to a related sandstone. Also of note is a map showing the distribution of red and gray shale colors.

Rich, J.L. Probable fondo origin of Marcellus–Ohio–New Albany–Chattanooga bituminous shales: Amer. Assoc. Petrol. Geol. Bull., v. 35, p. 2017–2040.

Classic analysis of origin of black shales by a famous early sedimentologist using sedimentary structures and physical stratigraphy, plus character of bounding units. First introduction by Rich of the terms *fondo*, *clino*, and *unda*, only the term *clino* having found wide acceptance.

Richter, R. 1931. Tierwelt und Umwelt im Hunsrückschiefer; zur Entstehung eines schwarzen Schlammsteins: Senckenbergiana, v. 13, p. 299–342.

An all-time classic by a famous early paleoecologist on the paleoecology of shale deposited in a tidal flat environment.

Rubey, W.W. 1931. Lithologic Studies of Fine-Grained Upper Cretaceous Sedimentary Rocks of the Black Hills Region. U.S. Geol. Survey Prof. Paper 165-A, 54 pp.

A classic study that used thin-section and chemical analyses plus fossil content and sedimentary structures to elucidate the origin of these shale formations, which total more than 1000 ft. thick. Among topics still of interest today are: (1) porosity and deformation: (2) fissility; and (3) bedding laminations (their origin and time significance). Author's plate 4 (not shown) has six photomicrographs of thin sections of shales. Today we still do not have too many studies comparable to this.

Ruedemann, R. 1935. Ecology of black mud shales of eastern New York: Jour. Paleontol., v. 9, p. 79–91.

A general review of the origin of black shales, plus a specific faunal comparison between the black Utica and the gray Lorraine Shales, the former being an impoverished bottom fauna and the latter a normal muddy-bottom fauna with many bryozoans, pelecypods, and gastropods. Additionally, black shales are divided into those with a purely planktonic versus those with a planktonic–benthonic fauna. Suggests deep, widespread basins for the majority of the Paleozoic black shales of the Appalachians.

Sorby, H.C. 1908. On the application of quantitative methods to the study of the structure and history of rocks: Quart. Geol. Soc. London, v. 64, p. 171–233.

Perceptive about shales far ahead of his time. See particularly pages 189–200, 211–212, and 220–224.

Strom, K.M. 1939. Land-locked waters and the deposition of black muds. *In:* P.D. Trask, Ed., Recent Marine Sediments. Amer. Assoc. Petrol. Geol., p. 356–372.

A condensed version of two papers which appeared in a Norwegian journal. Mainly an account of the hydrography of recent black mud environments (the author studied 30 fjords) and the generation of black mud. Seventy-five references of comparable deposits by German, Russian, French, Spanish, Japanese, American, Norwegian, and Swedish authors.

Woolnough, W.G. 1937. Sedimentation in barred basins and source rocks of oil: Amer. Assoc. Petrol. Geol. Bull., v. 21, p. 1101–1157.

Wide-ranging review by an experienced geologist with many ideas on argillaceous sedimentation in restricted basins. Classic and worth your time to see his points of view.

CLASSIFICATION

We have included but three papers on classification, a topic that nonetheless has long been very troublesome to almost all who work with shales. Additional viewpoints can be obtained by consulting the soil mechanics literature and some of the references in "Environmental and Engineering Geology"; almost always specific, practical needs entail specialized classifications.

• Picard, M.D. 1971. Classification of fine-grained sedimentary rocks: Jour. Sed. Petrol., v. 41, p. 179–195.

> Advocates a classification scheme, somewhat analogous to those of sandstones, using texture and mineralogy (optical and X-ray). For example, a typical description would consist of a textural term modified by a sandstone name and the dominant clay mineral.

Underwood, L.B. 1967. Classification and identification of shales: Proc. Am. Soc. Civil Engineers, v. 93, Paper 5560. Jour. Soil Mechanics and Foundations Div., n. SM6, p. 97–116.

> After a discussion of classification, geologic properties (grain size, mineralogy, fissility, etc.), significant engineering properties, such as modulus of elasticity, moisture, density, and void ratio and permeability are reviewed. Table 1 (not shown), an engineering evaluation of shales, is very useful; Table 2 (not shown) gives some physical properties of different shales that should be helpful as reference standards. See Fig. 1 (not shown) and also Fig 3.3.

Wood, L.E., and P. Deo. 1975. A suggested system for classifying shale materials for embankments: Assoc. Eng. Geol. Bull., v. 12, n. 1, p. 39–55.

> Proposes an operational classification of Indiana shales based on their slaking properties. A combination of slaking, slaking durability, and sodium sulfate soundness gives four categories of shale "suitability" for construction purposes.

TRANSPORT AND EROSION

The citations of this section emphasize the hydraulics of mud transport and erosion, the latter being intimately related to its physical properties at the sediment–water interface. Therefore, some of the citations are good sources for the effects of flocculation on settling velocity and for the effects of waves and tides on the transport and erosion of mud. Additionally, we have included a few references that emphasize paleocurrent systems—either indirectly from a dispersal map based on the areal variation

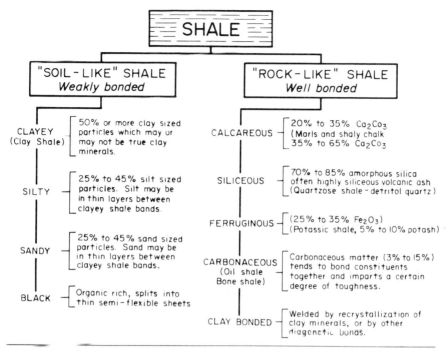

Figure 3.3 Classification of shale. (Redrawn from Underwood 1967, Fig. 3.)

of scalar properties or, much more rarely, from direct paleocurrent observations of fossil orientation or perhaps from primary sedimentary structures—or even concretions! Because the paleocurrent system of a sedimentary deposit—be it a widespread thin mud, a small fluvial sandstone
body, or a large carbonate basin—is one of the chief integrating factors in
making an interpretation, it deserves much more attention by all students
of shales. Holeman's (1968) paper was also included to give an insight
into the relative importance of big rivers as contributors of mud brought to
the world ocean.

See also the sections on "Deposition of Modern Muds" and "Mineralogy" for related papers.

Allersma, E. 1971. Mud on the oceanic shelf off Guyana. *In:* Symposium on Investigation and Resources of the Caribbean Sea and Adjacent Regions. UNESCO,
Paris, p. 193–203.

> Description of mud deposition along the coast downcurrent from the Amazon
> estuary. This deposit has less than 2% sand and shells and an average diameter
> of about 1μm. Transport is believed to be by a combination of waves and cur-
> rents which produce a series of migrating mud banks with a wavelength of about
> 45 km. These mud banks dampen waves near the shore and, thus create a type of
> positive feedback mechanism that helps entrap more coastal muds. One of the
> few studies of its kind, except for that of Nair (1976). Similar deposits should be
> sought in ancient rocks. See Fig. 3.4.

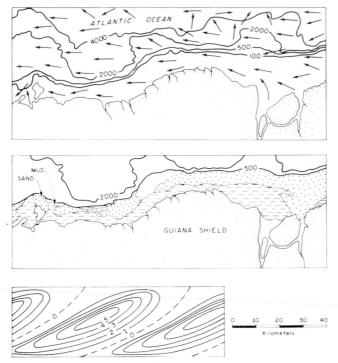

Figure 3.4 Coastal and oceanic circulation, distribution of mud and sand and "mega mud waves" off the Guiana Shield. Most of the mud is supplied by the Amazon River. (Redrawn from Allersma 1971, Figs. 1 and 2.)

Cole, R.D., and M.D. Picard. 1975. Primary and secondary sedimentary structures in oil shale and other fine-grained rocks, Green River Formation (Eocene), Utah and Colorado: Utah Geol., v. 2, p. 49–67.

>   Well illustrated and organized and includes a classification for the systematic description of structures in fine-grained argillaceous sediments, as well as estimates of their abundance in siltstone, claystone, sandstone, and carbonate. Well done.

Collins, M., G. Ferentinos, and F.T. Banner. 1979. The hydrodynamics and sedimentology of a high (tidal and wave) energy embayment (Swansea Bay, Northern Bristol Channel): Estuarine Coastal Mar. Sci. v. 8, p. 49–74.

>   Comprehensive study of current systems and bottom sediments shows how complicated it can be to relate the two in even a small area such as this—especially when relict sediments are recognized. Excellent illustrations.

Colton, G.W. 1967. Orientation of carbonate concretions in the Upper Devonian of New York. U.S. Geol. Survey Prof. Paper 575-B, p. 57–59.

>   Orientation of argillaceous carbonate concretions parallels paleocurrent structures, which, in turn, reflect primary depositional fabric in a study area about 110 miles long. This observation, which can be easily checked elsewhere,

suggests that elongate concretions may be a widely overlooked paleocurrent indicator in shales. See "Burial History" for more on concretions.

Chamley, H., and F. Picard. 1970. L'heritage détritique des fleuves provencaux en milieu marin: Tethys, v. 2, p. 211–226.

Clay mineralogy and heavy minerals of small streams that flow into the Mediterranean Sea. Compare with Quakernaat (1968).

Drake, D.E. 1976. Suspended sediment transport and mud deposition on continental shelves. In: D.J. Stanley and D.J.P. Swift, Eds., Marine Sediment Transport and Environmental Management. John Wiley and Sons, New York, p. 127–158.

Review of existing knowledge about suspended sediment: methods of study, hydrodynamics, mud sources, estuarine and shelf processes.

Einsele, G., R. Overbeck, H.U. Schwarz, and G. Unsöld. 1974. Mass physical • properties, sliding and erodibility of experimentally deposited and differentially consolidated clayey muds: Sedimentology, v. 21, p. 339–372.

Thorough, quantitative, experimental study of clay sedimentation and consolidation and subsequent erosion with a section on gravitational mass movements on slopes. Notes initial water contents up to 450% for unconsolidated clay. Stress on role of clay fabric. Good example of integration of different techniques. See also Migniot (1968).

Gibbs, Ronald J., Ed. 1974. Suspended Solids in Water. Plenum Press, New York • and London, 320 pp.

Nineteen papers in four parts (principles of suspension, techniques and nearshore and offshore studies). Basic reference essential for understanding the details of transport and deposition of mud. Also contains information on solution rates of slowly settling particles.

Griffin, G.M. 1962. Regional clay-mineral facies—Products of weathering intensity and current distribution in the northeastern Gulf of Mexico: Geol. Soc. Amer. Bull., v. 73, p. 737–767.

A provenance study based on clay mineralogy of modern marine and fluvial muds. Traces stream contributions through the littoral zone to the open shelf, with much emphasis on the variations in alluvial muds in the Mississippi River Basin. Over 600 modern samples, plus some Tertiary ones. Good methodologic model for the provenance study of of mud.

Hedges, J.I., and P.L. Parker. 1976. Land-derived organic matter in surface sediments from the Gulf of Mexico: Geochim. Cosmochim. Acta, v. 40, p. 1019–1029.

Shows that the proportion of terrestrial organic matter in fine-grained sediments can be traced by using C^{12}/C^{13} ratios and lignin derivatives (both higher for terrestrial organic matter). In modern sediments there seems to be very little terrestrial organic matter more than a few miles offshore, except near very large rivers. Ancient rocks need to also be studied using such provenance techniques, as well as phytoclasts. See Vandenberghe (1976).

Heezen, B.C., Ed. 1977. Influence of Abyssal Circulation on Sedimentary Accumulations in Space and Time. Developments in Sedimentology, No. 23, Elsevier Scientific Publ. Co., Amsterdam, 215 pp; also published as Mar. Geol., v. 23, 1977.

Eleven articles, mostly about modern and ancient paleocurrent (paleocirculation) systems in oceanic deeps. Of special interest is an article on abyssal bedforms and another on episodes of ocean stagnation. See also Ryan and Cita (1977).

Holeman, J.N. 1968. The sediment yield of major rivers of the world: Water Resources Res., v. 4, p. 737–747.

Still the basic reference for a worldwide summary of the amount of suspended load, mostly mud, carried to the world ocean, with Asia contributing the greatest share (see data in Holeman's Table 1).

Jones, M.L., and J. A. Clendening. 1968. A feasibility study for paleocurrent analysis in lutaceous Monongahela–Dunkard strata of the Appalachian Basin: Proc. West Virginia Acad. Sci., v. 40, p. 255–261.

Ostracods in many of the mudrocks and shales of the Monongahela and Dunkard strata are well oriented.

Jones, M.L., and J. M. Dennison. 1970. Oriented fossils as paleocurrent indicators in Paleozoic lutites of Southern Appalachians: Jour. Sed. Petrol., v. 40, p. 642–649.

Over 13 000 measurements of fossil orientation (graptolites, brachipods, cricocanarids, gastropods, ostracods, and tasmanites) in the Athens (Ordovician) and Chattanooga (Devonian) Shales. Authors suggest that whereas commonly hundreds of fossils must be measured at a locality to yield significant results, this is generally not a serious problem in shales because the fossils are small and usually a single bedding plane contains sufficient numbers. Five useful suggestions for additional research. Short but significant paper and a long-delayed follow-up to Ruedemann's (1897) study of graptolite orientation in Lower Paleozoic black shales in New York (Amer. Geol., v. 19, p. 367–391). See Fig. 3.5.

Kranck, K. 1975. Sediment deposition from flocculated suspensions: Sedimentology, v. 22, p. 111–123.

Grain size analysis of fine-grained sediments has been hampered by uncertainty about the relationship between measured distribution and originally flocculated particles. This paper suggests that most grains larger than the grain *mode* settle as single grains, those smaller as part of flocs. Could this relation be used in a grain-size study of shales? See Fig. 3.6.

Krone, R.B. 1962. Flume Studies of the Transport of Sediment in Estuarial Shoaling Processes. Hydraulic Eng. Lab. and Sanitary Eng. Res. Lab., Univ. California, Berkeley, 110 pp.

Rheologic study of clay sediments from San Francisco Bay to determine transportational and depositional properties. Flume experiments on naturally flocculated muds demonstrated flocculation processes and a rational theory for the development of migratory mud shoals in estuaries. Scour and sedimentation rates at various discharges are given.

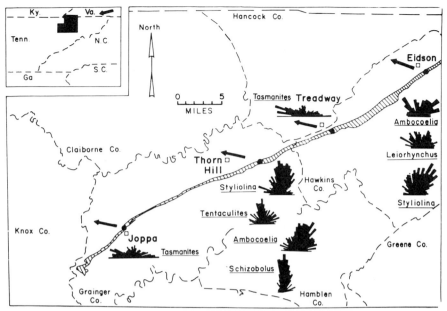

Figure 3.5 Fossil orientation in Chattanooga Shale along Clinch Mountain, Tennessee. (Jones and Dennison 1970, Fig. 6; published by permission of the authors and the Society of Economic Paleontologists and Mineralogists.)

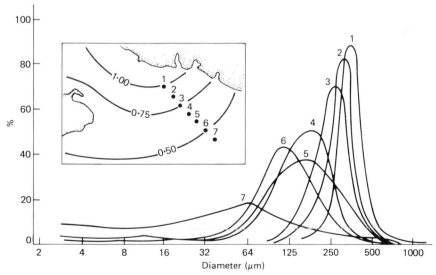

Figure 3.6 Grain size of bottom samples from an area of decreasing current speed at the south end of Abegweite Strait, Northumberland Strait. Contours give current speed in knots. (Krank 1975, Fig. 10; published by permission of the author and the International Association of Sedimentologists.)

Lonsdale, P., and Spiess, F. N. 1977. Abyssal bedforms explored with a deeply towed instrument package. *In:* B. C. Heezen, Ed., Influence of Abyssal Circulation on Sedimentary Accumulations in Space and Time. Developments in Sedimentology, No. 23, Elsevier Scientific Publ. Co., Amsterdam, p. 57–76; also published in Mar. Geol., v. 23, 1977.

> At water depths of 1.5–6 km, bottom photographs and side-looking sonar records reveal wave and current ripples, sand waves, mud waves, and erosional furrows. Depositional and erosional bedforms in cohesive sediment occur beneath the deepest thermohaline currents.

McCave, I. N. 1971. Wave effectiveness at the sea bed and its relationship to bedforms and deposition of mud: Jour. Sed. Petrol., v. 41, p. 89–96.

> Judging by the North Sea, mud deposition in a basin depends on suspended material concentration, near-bottom current velocity, and magnitude and frequency of wave activity. Believes that when mud concentration is high (greater than 100 mg/l), mud is deposited result independent of wave activity, but when it is low (less than 1 mg/l), wave activity is the controlling factor. Notes that there is generally an increase of turbidity near shore and cites conditions near Rhine, Po, and Amazon.

● McCave, I.N. 1972. Transport and escape of fine-grained sediment from shelf areas. *In:* D.J.P. Swift, D.B. Duane, and O.H. Pilkey, Eds., Shelf Sediment Transport: Process and Pattern. Dowden, Hutchinson and Ross, Stroudsburg, Pa., p. 225–248.

> Muddy shelves are classified into five types and it is found that mud in suspension decreases roughly exponentially offshore along plumes off deltas (areas of converging currents are an exception). Believes that most mud escaping from shelves is deposited on slope and rise. Critical erosion velocities are related to shear strength (consolidation and dewatering). See Figs. 3.7 and 3.8.

Migniot, C. 1968. Étude des propriétés physiques de différents sédiments très fins et de leur comportement sous des action hydrodynamiques. La Houille Blanche, v. 23, p. 591–620.

> A paper that goes far to summarize and explain the hydrodynamic and plastic properties of clays and is therefore relevant to both sedimentologists and engineering geologists. Fifty-seven figures. See Fig. 3.9.

Moors, H.T. 1969. The position of graptolites in turbidites: Sed. Geol., v. 3, p. 241–261.

> Orientation of graptolites in the pelitic part of a graded bed parallels paleocurrent directions from underlying sands and silts, suggesting that it may be possible to use interbedded siltstones to define paleocurrents in a thick shale sequence. Many references.

NEDECO. 1968. Surinam Transportation Study: Report on Hydraulic Investigation. Consultants Netherlands Engineering, The Hague, Netherlands, 293 pp.

> A comprehensive study in 13 chapters, the central theme of which is how to cope with over 100 million tons of mud transported yearly from the Amazon's

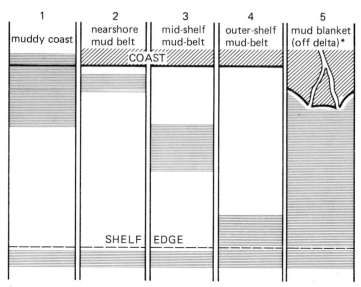

*or under advective mud stream

Figure 3.7 Schematic representation of five sites of shelf mud accumulation. (McCave 1972, Fig. 91; published by permission of the author and Dowden, Hutchinson, and Ross.)

mouth to the Guyana coast and into its estuaries. Many tabled hydraulic and sediment data on muddy estuaries and on tidal processes, plus deposition and erosion of mud along the coast. Notable for the discussion of heavy silt suspensions called "sling mud." *Key words:* Coastal geomorphology and erosion, coastal mud belt, migrating mud banks, plus coastal and tidal hydraulics. See also Allersma (1971).

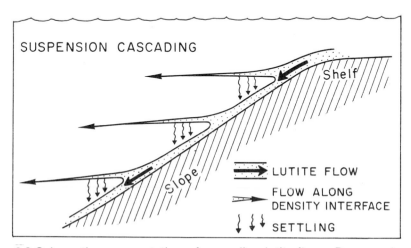

Figure 3.8 Schematic representation of cascading lutite flows. Decrease in width of arrows indicates decrease in concentration. (McCave 1972, Fig. 96; published by permission of the author and Dowden, Hutchinson and Ross.)

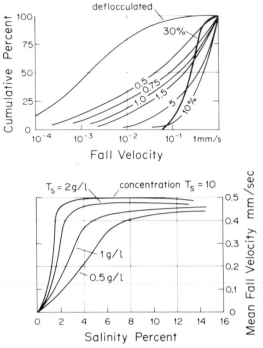

Figure 3.9 Effect of salinity on flocculation and fall velocity. (Redrawn from Migniot 1968, Fig. 9.)

Rashid, M.A., and G.E. Reinson. 1979. Organic matter in surficial sediments of the Miramichi Estuary, New Brunswick, Canada: Estuarine Coastal Mar. Sci., v. 8, p. 23–36.

A possible model for a comparable study of an ancient shale, the key words being: well-mixed estuarine waters; estuarine sedimentation, organic matter, sediment size, isotopic composition, circulation processes, carbon, and nitrogen, and gas chromatography.

Southard, J.B. 1974. Erodibility of fine abyssal sediment. *In:* A.L. Inderbitzen, Ed. Deep-Sea Sediments. Plenum Press, New York and London, p. 367–379.

An essay review paper with some experimental data; suggests four important factors determine erodibility: dynamics of turbulent boundary layer; physical–chemical nature of sediment; past history, including rate of sedimentation and degree of flocculation; and activity of benthic organisms. Eighteen references.

Stow, D.A.V., and J.P.B. Lovell. 1979. Contourites. Their recognition in modern and ancient sediments: Earth-Sci. Rev., v. 14, p. 251–291.

Identifies muddy and sandy contourites. Muddy contourites are bioturbated and may contain biogenic sand concentrated in irregular layers. They have relatively high contents of $CaCO_3$ and organic carbon compared with associated turbidites. They differ from very fine-grained turbidites in the absence of a sequence of sedimentary structures. In sandy contourities, a 90° divergence in paleocurrents differentiates turbidites and contourites in the same sequence.

·• Swift, D.J.P., J.R. Schubel, and R.W. Sheldon. 1972. Size analysis of fine-grained suspended sediments: A review: Jour Sed. Petrol. v. 42, p. 122–134.

Essentially four methods of size distribution analysis of suspended particles are described: microscopic observation; sedimentation analysis; optical analysis (usually light scattering); and Coulter counter (electronic) analysis. The Coulter counter appears to be, overall, the most satisfactory approach.

Tomadin. L. 1974. Les minéraux argileux dans less sediments actuels de la mer Tyrrhenienne: Bull. Groupe Francais Argiles, v. 26, p. 219–228.

Dispersal map of 112 samples of modern mud from a small sea, combined with schematic maps of actual currents at surface, intermediate, and deep levels.

Vandenberghe, N. 1976. Phytoclasts as provenance indicators in the Belgian Septaria Clay of Boom (Rupelian age): Sedimentology, v. 23, p. 141–145.

Measurement of the reflectivities of phytoclasts and coal particles of silt size can be useful as provenance indicators in muds and shales. See Bostick and Foster (1973) for use of this measurement in reconstructing thermal history.

van Straaten, L.M.J.U., and Ph. H. Kuenen. 1957. Accumulation of fine-grained sediments in the Dutch Wadden Sea: Geol. Mijnbouw, N.W. series, v. 19e, p. 329–354.

Fine-grained sediment, derived from the North Sea, is trapped in the estuaries of Holland by tidal currents that are competent to transport the suspended solids on the flood tide, but not to erode them on the ebb. Therefore, the North Sea is being winnowed of its fine-grained fraction, which is moved to the shore, resulting in a facies separation.

Venkatarathnam, K., P.E. Biscaye, and W.B.F. Ryan. 1972. Origin and dispersal of Holocene sediments in the eastern Mediterranean Sea. In: D.J. Stanley, Ed., The Mediterranean Sea. Dowden, Hutchinson and Ross, Stroudsburg, Pa., p. 459–469.

Uses factor analysis of trace element concentration and clay mineralogy to sepa-
rate source areas of fine-grained sediments. Five source areas were identified, including a carbonate-rich, wind-blown contribution from North Africa.

Verger, F. 1969, Marais et Wadden du Littoral Francais. Biscaye Frères, Bordeaux, 541 pp.

Brief section on tidal dynamics followed by two major parts: wadden and marais, the former with 276 pp. and 13 chapters and the latter with 143 pp. and 11 chapters. In essence, a comprehensive regional description that encom-
passes geomorphology, pedology, sedimentary processes, and resultant deposits and their historical development all along the long Atlantic coastline of France. Many line drawings and aerial photographs, 231 in all, plus some compositional data on mineralogy, fauna, and vegetation, and a number of historical maps.

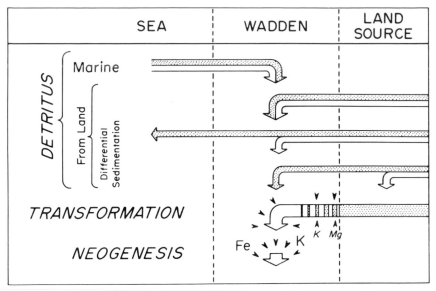

Figure 3.10 Multiple sources of coastal mud and chemical transformation along muddy shores of the Atlantic coast of France. (Redrawn from Verger 1969, Fig. 29.)

Block diagrams of mechanics of mud deposition, rills, and rill networks are notable as is a plate at the scale of 1:10 000 of the tidal zone at l'Anse de l'Aiguillon, a remarkable geomorphic map. Over 700 references, plus separate indices for aerial photographs and definitions. If you are studying an argillaceous deposit that you suspect is proximal to a tidally dominated shoreline, be sure to examine this major source book. See Fig. 3.10.

Wells, J.T., and J.M. Coleman. 1978. Longshore transport of mud by waves: Northeastern coast of South America: Geologie en Mijnbouw, v. 57, p. 353–359.

Time series measurements of water waves and study of migrating mudwaves suggest that as much as 70×10^6 cubic meters of mud may be transported yearly. See also Allersma (1971).

Wiley, M., Ed. 1976. Estuarine Processes, v. II. Circulation, Sediments, and Transfer of Material in the Estuary. Academic Press, Inc., New York, 428 pp.

Twenty-seven papers with eleven on sedimentation, including discussion of the Amazon, Gironde, San Francisco Bay and Chesapeake Bay plus others on organic matter, nutrients, hydraulics and circulation.

Young, R.N., and J.B. Southard. 1978. Erosion of Fine-Grained Marine Sediments: Sea-Floor and Laboratory Experiments: Geol. Soc. Amer. Bull., v. 89, p. 663–672.

Observations at sea made with a sea-floor flume revealed critical shear velocities of 0.32 to 0.84 cm/s; tidal currents had similar values. Tabled data plus an appendix showing how shear velocity is obtained from threshold surface velocity.

DEPOSITION OF MODERN MUDS

Here we have included chiefly studies of mud deposition in modern marine basins, coastal as well as deep ocean. In contrast to many of the other parts of this bibliography, there are a number of books and monographs that describe, commonly with good integration, specific basins.   Perhaps because of their ease of study, it seems to us that sedimentologists have worked on modern muds more than on their ancient equivalents, especially since the great post-World War II boom in marine geology. Mud accumulations have been documented from a number of deep basins, from deltaic deposits, from estuaries and tidal flats, in lakes, and from open shelves off the mouths of big rivers, such as the Amazon and the Po.

Attia, M.I. 1954. Deposits in the Nile Valley and the Delta. Geol. Survey Egypt, Cairo, 356 pp.

> A summary of the borings along 600 km in the lower Nile, with many transverse and several longitudinal sections. Basically a reference collection of the logs of these borings but also useful information of general interest. For instance, the average thickness of the top alluvial clays (Nile mud) is 8.40 m. Seven chapters, the last being a short summary.

Bouma, A.H. 1963. A graphic presentation of the facies model of salt marsh deposits: Sedimentology, v. 2, p. 122–129.

> A brief but useful summary of some of the sedimentary structures, trace fossils, and other features that can be used to distinguish fine-grained deposits formed in salt marshes.

Chamley, H. 1971. Recherches sur la Sédimentation Argileuse en Mediterranée: Sci. Géol. Mem., v. 35, 209 pp.

> Four chapters covering present and Quaternary argillaceous sedimentation, paleoclimatology, organic- versus detrital-rich clays, chemical sedimentation, contributions of stream material, and methods. Actual work is based on only three subregions of the Mediterranean, but comprehensive literature coverage (almost 500 references) reinforces generality of conclusions. Figure 50 (not shown) gives model for Mediterranean. Rich source of data and very useful as model study.

Clague, J.J. 1976. Sedimentology and Geochemistry of Marine Sediments Near Comox, British Columbia. Geol. Survey Canada Paper 76-21, 21 pp.

> Clayey and sandy silt in an estuary, derived from both rivers and the erosion of coastal bluffs, are studied both geochemically and texturally. Good integration of seismic profiles, hydrography, geochemistry, and sedimentology, and there is an interesting sediment dispersal map (Fig. 3.11). Base metals are highest in finest fraction.

Coleman, J.M. 1966a. Ecological changes in a massive fresh-water clay sequence: Trans. Gulf Coast Assoc. Geol. Socs., v. 16, p. 159–174.

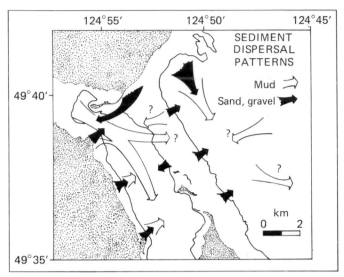

Figure 3.11 Generalized sediment dispersal: transport to sinks by rivers, currents and waves. (Clague 1976, Fig. 11.)

Excellent paper on deposition and diagenesis of fine-grained sediments in freshwater environments. A seemingly homogeneous clay–silt sequence was subdivided, using X-ray radiographs, into five facies: poorly drained swamp, well-drained swamp, lacustrine, lacustrine delta, and channel fill. The most prominent diagenetic feature is the development of concretions of various iron minerals, including pyrite, vivianite, and siderite.

Coleman, J.M. 1966b. Recent Coastal Sedimentation: Central Louisiana Coast. Coastal Studies Ser., v. 17, Louisiana State Univ. Press, Baton Rouge, 73 pp.

Maps surface sedimentary environments west of the mouth of the Mississippi and describes their recent sediments, emphasizing their lithology, sedimentary structures, flora, and fauna. Mud is the dominant sediment. Well illustrated.

Degens, E.T., and D.A. Ross, Eds. 1974. The Black Sea—Geology, Chemistry and Biology. Amer. Assoc. Petrol. Geol. Mem. 20, 633 pp.

Five sections—"Structure," "Water," "Sediments," "Biology," and "Geochemistry"—plus a short summary by K.O. Emery and J.M. Hunt give a very detailed overview of this often cited modern analog for some ancient black shales. Forty-three articles.

• Deuser, W.G. 1975. Reducing environments. *In:* J.P. Riley and G. Skirrow, Eds., Chemical Oceanography, v. 3. 2nd ed. Academic Press, New York, p. 1–37.

An attempt to marshal evidence from physical oceanography concerning modern reducing environments in the open ocean. A good introduction to sedimentary sequences in which reducing conditions are not caused by restriction of circulation. See Fig. 3.12.

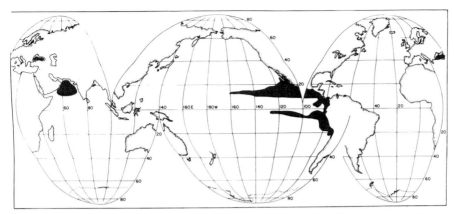

Figure 3.12 World distribution of oxygen-deficient (less than 20 μg-at/l) intermediate and deep water masses as compiled from various sources. (Deuser 1975, Fig. 16.1; published by permission of the author and the University of Chicago, Department of Geography.)

• Eden, W.J. 1955. A laboratory study of varved clay from Steep Rock Lake, Ontario: Amer. Jour. Sci., v. 253, p. 659–674.

  Variations of grain size, water content, and plasticity within varves.

Eisma, D., and H.W. van der Marel. 1971. Marine muds along the Guyana coast and their origin from the Amazon Basin: Contrib. Mineral. Petrol., v. 31, p. 321–334.

  A provenance study of a stream of mud 20–40 km wide, which moves along the Guyana coast carrying a load of about 1 to 2 × 10⁸ tons annually. These suspended muds have virtually the same composition as those of the Amazon and are very different from the composition of mud in the Guyanese soils and rivers. The only measurable effect of sea water on the mud particles traveling from the mouth of the Amazon to Venezuela is some decrease in potassium fixation. See Parham (1966) for the clay mineral zonation found in many ancient basins, Allersma (1971) and Nedeco (1968) for local geology, and Gibbs (1977) for further discussion of mineralogy.

Emery, K.O. 1969. Distribution pattern of sediments on the continental shelves of western Indonesia: Economic Commission for Asia and the Far East, Technical Bull., v. 2, p. 79–82.

  The shelf north of Indonesia is one of the world's largest areas of shale deposition and probably our closest analog to the wide, shallow seas of much of the geologic past. A map shows the area to be covered mostly by mud, with some fringing coral reefs and patches of relict sand. See Hamilton's (1974) map of Indonesian sedimentary basins and his description of the role of clay and shale in their deformation.

Endyanov, E.M. 1972. Principal types of recent bottom sediments in the Mediterranean Sea: Their mineralogy and geochemistry. In: D.J. Stanley, Ed., The Mediterranean Sea. Dowden, Hutchinson and Ross, Stroudsburg, Pa., p. 355–399.

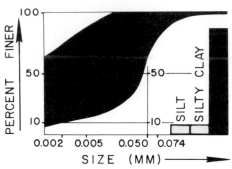

Figure 3.13 Size distribution of mud in the alluvial valley of the Missouri River. (Redrawn from Glenn and Dahl 1959, Table 1.)

Summary of the distribution of chemical and some mineralogic properties. The Mediterranean is a good place to study the modern-day relation of mud to other sediment types.

Evans, G. 1965. Intertidal flat sediments and their environments of deposition in the Wash: Quart. Jour. Geol. Soc. London, v. 121, p. 209–245.

Excellent study of the sedimentology and biology of the various subenvironments in this depositional setting. Clays and silty clays in this example are confined to the salt marsh.

Fisk, H.N. 1947. Fine-Grained Alluvial Deposits and Their Effects on Mississippi River Activity. U.S. Army Corps of Engineers, Mississippi River Commission, Vicksburg, Miss., 82 pp.

Fisk divided the fine-grained alluvium of the Mississippi Valley into meander belt, back swamp, braided stream, and delta-plain deposits. Then the engineering properties of each type of deposit were described. Still an excellent source of information on the way fine-grained material accumulates in alluvial environments. Well illustrated. Compare with Coleman (1966).

Gibbs, R.J. 1973. The bottom sediments of the Amazon shelf and tropical Atlantic Ocean: Mar. Geol., v. 14, p. M39–M45.

One of the very few regional studies—this one about an elongate, muddy shelf almost 2000 km long—of variation in composition and grain size of a modern mud. See also Heath and others (1974).

Glenn, J.L., and A.R. Dahl. 1959. Characteristics and distribution of some Missouri River deposits: Proc. Iowa Acad. Sci., v. 66, p. 302–311.

Environment of alluvium controls its distribution and physical properties. Figure 3.13 shows the particle size distribution of 43 samples from each of 13 channel fills in the floodplain of the Missouri River.

Gopinathan, C.K., and S.Z. Qasim. 1974. Mud banks of Kerala—Their formation and characteristics: Indian Jour. Mar. Sci., v. 3, p. 105–114.

Description of the dynamics of mud bank formation using echograms, soundings, and sequential observations over several years. Turbid suspensions proximal to bank dampen wave energy. See also Jacob and Qasim (1974).

Gucluer, S.M., and M.G. Gross. 1964. Recent marine sediments in Saanich Inlet, a stagnant marine basin: Limnol. Oceanog., v. 9, p. 359–376.

Comprehensive study of bedding, reducing capacity, and nitrogen, carbon, and carbonate contents, plus microfauna, of the poorly oxygenated sediments of a small silled basin. Five sediment sources distinguished, one being the effluent from a cement plant.

Heath, G.R., T.C. Moore Jr., and G.L. Roberts. 1974. Mineralogy of surface sediments from the Panama Basin, eastern equatorial Pacific: Jour. Geol., v. 82, p. 145–160.

Fines dispersed primarily by surface currents and, to a lesser extent, by bottom currents, with winnowing having little effect. Clay mineral patterns and heavy minerals indicate shoreline proximity. Compare with Gibbs (1973).

Jacob, P.G., and S.Z. Qasim. 1974. Mud of a mud bank in Kerala, south-west coast of India: Indian Jour. Mar. Sci., v. 3, p. 115–119.

A description of the mud in the bank includes its particle size, phosphorus content, organic carbon, chlorophyll a, and phaeophytin contents and carbohydrates. The mud banks seem to increase fertility of the waters for fishing.

Kolb, C.R., and R.I. Kaufman. 1967. Prodelta clays of southeast Louisiana. In: A.F. Richards, Ed., Marine Geotechnique. Internatl. Marine Geotechnical Res. Conf. (Monticello, Illinois) Proc., Univ. Illinois Press, Urbana, p. 3–21.

Bedding properties, as revealed by X-ray radiographs, and some engineering properties were measured on cores. Four delta lobes were identified. Each deposit begins with slow sedimentation and equilibrium compaction, but as the sedimentation rate increases with the continuation of progradation, the clays become undercompacted and flowage, fracturing, and the development of slickensides become common. See Fig. 3.14.

Laking, P.N. 1974. The Black Sea, Its Geology, Chemistry, Biology: A Bibliography. Woods Hole Oceanographic Institution, Woods Hole, Massachusetts, 368 pp.

Over 4000 references to this fascinating sea from Russian, Romanian, Bulgarian, Turkish, and Western European language publications. Titles from 1681 to mid-1973, including Degens and Ross (1974). Helpful subject index.

Lisitzin, A.P. 1972. Sedimentation in the World Ocean. Soc. Econ. Paleontol. Mineral. Spec. Publ. No. 17, 218 pp.

The best single source for dispersion, in air and water, of land-derived material in the world ocean, mostly, but not entirely, using Russian sources. Much about deep marine mud.

Melieres, F., and H. Perez-Nieto. 1973. Les mineraux argileux des sediments Récents du Golfe de Cariaco (Vénézuela): Bull. Groupe Francais Argiles, v. 25, p. 65–78.

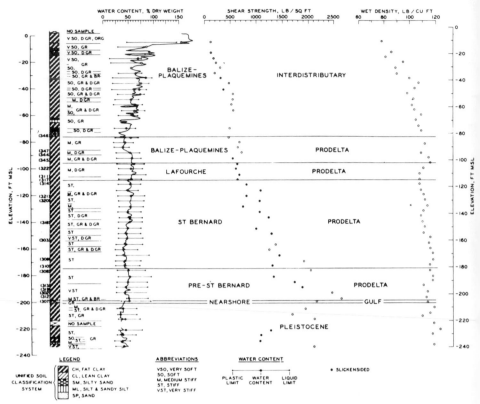

Figure 3.14 Graphic log, engineering properties, and selected radiography of cores from a boring in the alluvial valley of the Mississippi River. (Kolb and Kaufman 1967, Fig. 6; published by permission of the authors and the University of Illinois Press.)

Systematic maps of clay mineral composition (43 samples) in a narrow, 60 × 15 km gulf, the current system of which is known. Authors emphasize the role of differential sedimentation. Because the clay mineral suite is mostly illite–kaolinite, variation in the crystallinity proved to be helpful in discriminating sources.

Monaco, A. 1975. Les facteurs de la sédimentation marine argileuse. Les phénomènes physico-chimiques á l'interface: Bull. B.R.G.M., 2nd Ser., Sec. IV, p. 147–174.

Study of modern muds off the Rhone Delta, east of Gibraltar, and in the straits between Sicily and Tunisia yields a cross-sectional model of mud transport offshore (Fig. 15, not shown), which should be of general interest.

Moore, D.G., J.R. Curray, and F.J. Emmel. 1976. Large submarine slide (olistostrome) associated with Sunda Arc subduction zone, northeast Indian Ocean: Mar. Geol., v. 21, p. 211–226.

A slide in Bengal Fan turbidites, covering 900 km³ generated a mud flow that spilled into a tributary channel on the fan and filled it for 145 km. Interesting mechanism for producing a mud-filled channel.

Nair, R.R. 1976. Unique mud banks, Kerala, southwest India: Amer. Assoc. Petrol. Geol. Bull., v. 60, p. 616–621.

One of the few descriptions of modern terrigenous mud banks that occur along open coasts. These consist of organic-rich fine silt and clay and occur in ox-ygen-deficient bottom waters (only about 10 m deep). The muds are thixo-tropic, appear to be gas generators, and have some associated mud eruptions. The banks and their associated suspended muds spectacularly damp wave ac-tivity along the shore. See also Allersma (1971), Hedberg (1974), Gopinathan and Qasim (1974), and Jacob and Qasim (1974).

Neiheisel, J. 1966. Significance of Clay Minerals in Shoaling Problems. Committee Tidal Hydraulics, U.S. Army Corps of Engineers, Tech. Bull. 10, 30 pp.

Clay minerals studied in modern sediments to help define shoaling in har-bors—in short, a provenance study using the X-ray and clay minerals rather than the petrographic microscope and sand grains.

Paul, J. 1970. Sedimentologische Untersuchungen eines küstennahen mediterranen Schlammbodens (Limski Kanal, Nördliche Adria): Geol. Rundschau, v. 60, p. 205–222.

A long, drowned river valley on the Yugoslavian Adriatic coast, dominantly filled with terra rosa mud and introduced detrital carbonate debris, was studied: grain size, water, carbonate, Fe and Mn contents, as well as vertical profiles of Eh and pH. English, French, and Russian abstracts.

Pelletier, B.R., F.J.E. Wagner, and A.C. Grant. 1968. Marine geology. In: C.S. Beals, Ed., Science, History and Hudson Bay, v. 2, Canada Dept. Energy, Mines and Resources, p. 557–613.

This modern sediment study describes recent sedimentation in Hudson Bay, emphasizing terrigeneous dispersal and present current systems. Modern clays accumulate in deeper parts of basin and in some sharply defined synclines of Belcher Islands. Reddish brown color in center of Hudson Bay is believed to be a response to slower sedimentation. Fairly good regional model of clay deposi-tion in a wide, shallow sea. See Fig. 3.15.

Pevear, D.R. 1972. Source of Recent nearshore marine clays, southeastern United States. In: B. Nelson, Ed., Environmental Framework of Coastal Plain Estuaries. Geol. Soc. Amer. Mem., v. 133, p. 317–336.

Clay minerals of rivers of the southeastern United States reflect piedmont weathered soils and contain chiefly kaolinite, plus some vermiculite and smec-tite. However, nearshore clays are mostly smectite, suggesting derivation from the shelf instead of the rivers. Good example of systematic dispersion and provenance study along a coastline.

Pilkey, O.H., and D. Noble. 1966. Carbonate and clay mineralogy of the Persian Gulf: Deep-Sea Res., v. 13, p. 1–16.

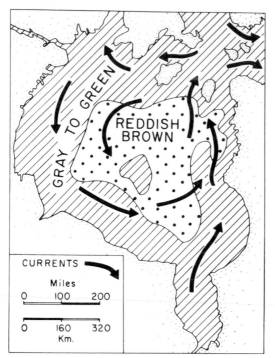

Figure 3.15 Generalized surface circulation and color of bottom muds of Hudson Bay. (Redrawn from Pelletier 1968, Figs. 11 and 15.)

Forty-five samples used to map clay and carbonate minerals in this large gulf show trends roughly parallel to its axis. There are surprisingly few studies like this.

Piskin, K., and R.E. Bergstrom. 1975. Glacial Drift in Illinois: Thickness and Character. Illinois Geol. Survey Circ. 490, 35 pp.

The stratigraphy and character of Pleistocene glacial drift, ranging from a few to almost 200 m deep, in Illinois are described. The tills and loess of these unconsolidated deposits are clay rich, are the parent materials for most Illinois soils, are important sources of building products, and have an effect on land use, mining, construction, and drilling. Excellent overview and a good reference for those studying ancient glacial deposits.

Plumb, C.E., G. Ricks, D.C. McMullen and C. Chastain. 1955. Seepage Conditions in Sacramento Valley. California Div. Water Resources, Rept. to the Water Project Authority, 128 pp.

Detailed study of the hydrology of various types of floodplain deposits in the Sacramento Valley. Sixty-eight maps, cross sections, and hydrographs. Contains grain-size analyses of each type of deposit and qualitative data on permeability, but no analyses. Fine-grained deposits are most common in the downstream portion of abandoned meander loops (although rarely filling the entire channel) and in "flood basins" in the lower part of the valley.

Pryor, W.A. 1975. Biogenic sedimentation and alteration of argillaceous sediments in shallow marine environments: Geol. Soc. Amer. Bull., v. 86, p. 1244–1254.

An interesting and unusual paper that calls attention to the possible general importance of organisms as producers of pelletal muds and their contribution to glauconite. How often do clay mineralogists consider penecontemporaneous clay mineral alteration by the lower intestine?

Reineck, H.-E., W.F. Gutmann, ünd G. Hertweck. 1967. Das Schlickgebiet südlich Helgoland als Beispiel rezenter Schelfablagerungen: Senckenbergiana Lethaea, v. 48, p. 219–275.

Sediment types, fabrics, shells, and bioturbation facies of a muddy shelf area in front of two river mouths are studied and an attempt is made to correlate facies with hydrographic data. Strong emphasis on bioturbation. Thin sand layers, some of which can be correlated for up to 15 km, are believed to be the result of storms. See Fig. 3.16. Well-illustrated English abstract and summary.

Reineck, H.-E., and I.B. Singh. 1973. Depositional Sedimentary Environments. Springer-Verlag, New York, Heidelberg, Berlin, 439 pp.

This book is very richly illustrated and contains much on the modern, terrigenous depositional environments of near-shore marine mud, plus much on the general processes of mud deposition and transport, as well as its early history of bioturbation and inorganic sediment structures. Almost 900 references, many to mud. See Figs. 3.17–3.19.

Rupke, N.A. 1975. Deposition of fine-grained sediments in the abyssal environments of the Algéro Balearic Basin, western Mediterranean Sea: Sedimentology, v. 22, p. 95–109.

Muds of hemipelagic and turbidity current origin are about equal contributors to these sediments. The turbidite muds distinguishable by grading, less bioturbation, and absence of sand-sized microfossils. Clay mineralogy may be distinctive. Hemipelagic muds show little variation vertically or laterally in the basin. Compare, please, Hesse (1975).

Rusnak, G.A. 1960. Sediments of Laguna Madre, Texas. In: F.P. Shepard, F.B. Phleger, and Tj. H. van Andel, Eds., Recent Sediments, Northwestern Gulf of Mexico. Amer. Assoc. Petrol. Geol. Tulsa, Okla., p. 153–196.

Sedimentologic study of the sedimentary fill of Laguna Madre. Includes bathymetry, hydrography, lithology, sedimentary structures, sediment colors, faunal distribution, and chemistry. Compares the subhumid northern part of Laguna Madre to the semiarid southern part. See Figs. 3.20 and 3.21.

Ryan, W.B.F., and M.B. Cita. 1977. Ignorance concerning episodes of ocean-wide stagnation. In: B.C. Heezen, Ed., Influence of Abyssal Circulation on Sedimentary Accumulations in Space and Time. Developments in Sedimentology, No. 23, Elsevier Scientific Publ. Co., Amsterdam, p. 197–215; Also published in Mar. Geol., v. 23, 1977.

Discussion of both the evidence and reasons for basin-wide abyssal stagnation

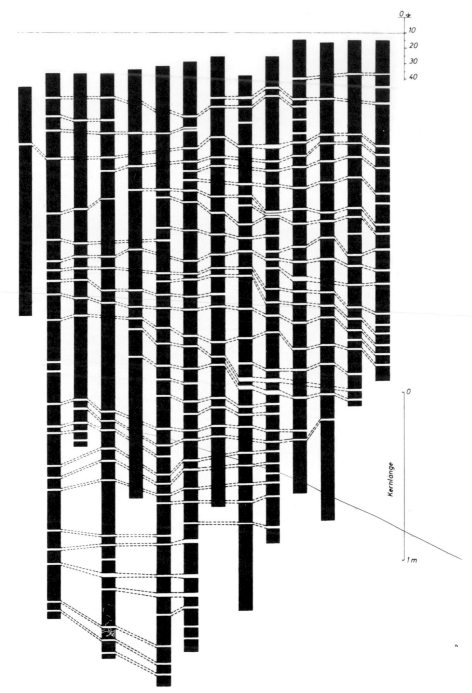

Figure 3.16 Correlation of fine sandy laminations (white) in mud (black) over a distance of 15 km. (Reineck and others 1967, Fig. 5.)

Figure 3.17 Block diagram of mud with lenticular, isolated sand and silt lenses. (Reineck and Singh 1973, Fig. 171; published by permission of the authors and Springer-Verlag.)

in the Mediterranean Sea (Pleistocene) and in the Atlantic and Indian Oceans (Cretaceous). A must for all students of organic-rich shales.

Schäfer, W. 1956. Wirkungen der Benthos-Organismen auf den jungen Schichtverband: Senckenbergiana Lethaea, v. 37, p. 183–263.

Key ideas include: bioturbation in interbedded modern mud, silt, and sand, experiments, and functional behavior of organisms. Pages 254–259 summarize most of the specific results of this study and provide an excellent model for others to follow.

Sinha, E., and B. McCosh. 1974. Coastal-Estuarine and Near-Shore Processes. Prepared for the Office of Water Resources Research by Ocean Engineering Information Service, La Jolla, Ca., 218 pp. (Distributed by National Technical Information Service, U.S. Dept. of Commerce).

A very complete annotated bibliography with 1009 references. Useful cross indices by subject (eight major headings) and by geographic area plus a listing of earlier bibliographies that contained 50 or more references.

• Stanley, D.J., and D.J.P. Swift, Eds. 1976. Marine Sediment Transport and Environmental Management. John Wiley and Sons, New York, 602 pp.

Twenty chapters covering almost all aspects ranging from ecology to facies to hydraulics of mud deposition. Very well illustrated (Fig. 3.22) and much tabled data (Table 3.1)

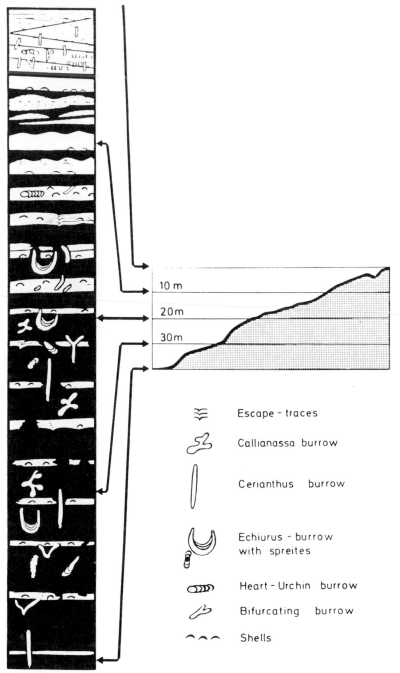

Figure 3.18 Schematic representation of deposits of various environments: shelf mud, transition zone, and coastal sand deposits, and their relation to the water depth. (Reineck and Singh 1973, Fig. 485; published by permission of the authors and Springer-Verlag.)

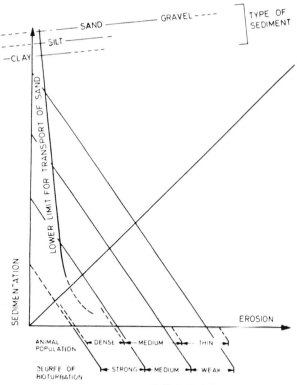

Figure 3.19 Relationships of grain size, animal population and bioturbation to the processes of erosion and sedimentation. The increasing rate of erosion causes coarsening of sediment, a smaller animal population and less bioturbation. (Reineck and Singh 1973, Fig. 227; published by permission of the authors and Springer-Verlag.)

Stuart, C.J., and C.A. Caughey. 1976. Form and composition of Mississippi Fan: Trans. Gulf Coast Assoc. Geol. Socs., v. 24, p. 333–343.

A description of the mud-rich Mississippi submarine fan, which is nicely summarized in Table 2 (not shown). The characteristics of this fan should also apply to others that have accumulated in almost closed, tectonically quiescent basins.

van Andel, Tj. H. 1964. Recent marine sediments of the Gulf of California. *In:* Tj. H. van Andel and G.G. Shor, Eds., Marine Geology of the Gulf of California. Amer. Assoc. Petrol. Geol. p. 210–310.

This small sea is flanked by nearby high-relief source areas and has a major delta at its closed end. Pure clays are virtually absent, but silty clays cover most of the gulf, except for sands near the Colorado Delta and very close to the shoreline. These muds will produce shales with interbedded chert and limestone and a distinctive facies relation with the sands.

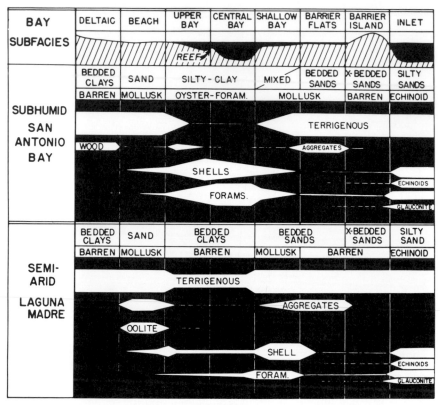

BAY	DELTAIC	BEACH	UPPER BAY	CENTRAL BAY	SHALLOW BAY	BARRIER FLATS	BARRIER ISLAND	INLET
SUBFACIES				REEF				

SUBHUMID SAN ANTONIO BAY	BEDDED CLAYS	SAND	SILTY - CLAY		MIXED	BEDDED SANDS	X-BEDDED SANDS	SILTY SANDS
	BARREN	MOLLUSK	OYSTER-FORAM.			MOLLUSK	BARREN	ECHINOID
						TERRIGENOUS		
	WOOD				AGGREGATES			
			SHELLS					
			FORAMS.					ECHINOIDS
								GLAUCONITE

SEMI-ARID LAGUNA MADRE	BEDDED CLAYS	SAND	BEDDED CLAYS		BEDDED SANDS		X-BEDDED SANDS	SILTY SAND
	BARREN	MOLLUSK	BARREN		MOLLUSK		BARREN	ECHINOID
			TERRIGENOUS					
					AGGREGATES			
	OOLITE							
				SHELL				ECHINOIDS
			FORAM.					GLAUCONITE

Figure 3.20 Diagrammatic representation of structural, textural, and compositional features which characterize the bay facies of Laguna Madre. (Rusnak 1960, Fig. 30; published by permission of the Society of Economic Paleontologists and Mineralogists).

van Andel, Tj. H. 1967. The Orinoco Delta: Jour Sed. Petrol., v. 37, p. 297–310.

Description of a depositional area receiving large amounts of fine-grained sediment that is extensively reworked by strong longitudinal current systems. Facies maps show contrasts with other deltas.

van Straaten, L.M.J.U. 1965. Sedimentation in the northwestern part of the Adriatic Sea. In: W.F. Whitland and R. Bradshaw, Eds., Submarine Geology and Geophysics. Butterworths, London, p. 143–162.

The Po River is depositing a mud blanket on the Italian side of the Adriatic. This argillaceous deposit has the form of a terrace, 2–20 km wide at a depth of 22–25 m. The terrace edge shows clinoform bedding. The molluscan fauna was extensively investigated and found to increase in diversity away from areas of greatest mud accumulation. The greatest mollusk abundance is very near shore, but the highest shell abundance is in offshore relict sands beyond the mud terrace, because there is so little terrigenous supply. See Fig. 3.23.

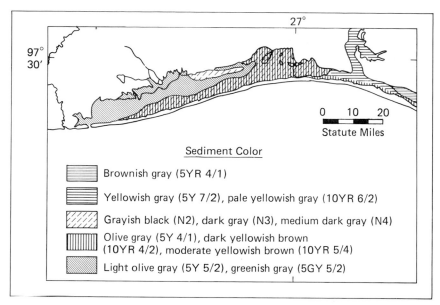

Sediment Color

Figure 3.21 Sediment colors in Laguna Madre: areas of algal mat impose a dark gray overtone to the basic sediment color and are illustrated by a double pattern. (Rusnak 1960, Fig. 24, published by permission of the Society of Economic Paleontologists and Mineralogists.)

MINERALOGY

Mineralogic studies of shales have been confined, with very few exceptions, to clay mineralogy. Over the past 10 years we have learned that clays do not show appreciable chemical transformations in response to their environment of deposition, and so are much better indicators of source than of sink. There is, however, a physical response to salinity changes in the depositional basin: differential flocculation causing more kaolinite near shore and smectite–illite offshore in recent sediments. More studies of this pattern in ancient sediments are needed. Unfortunately, diagenesis reduces the amount of provenance information in ancient shales, primarily because of the conversion of smectite to illite and other clays.

We know of few regional studies of the nonclay minerals in shales, except for zeolites in lake beds, other than those by Pelzer (1966) and Davis (1970). Surely, however, much information is contained in the nonclay mineralogy of shales.

Amiri-Garroussi, K. 1977. Origin of montmorillonite in Early Jurassic shales of NW Scotland: Geol. Mag., v. 114, p. 281–290.

Clay mineralogy is used to show a change in the provenance of the Broadford Shale from a basic igneous rock for the Lower Broadford to a metamorphic–sedimentary terrain for the Upper Broadford.

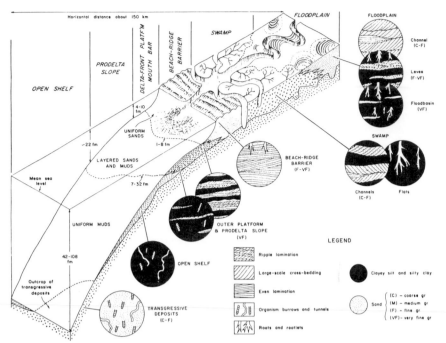

Figure 3.22 Depositional environments and sedimentary facies of the Niger Delta and Niger shelf: progressive size sorting of sediment results in a decrease in grain size through successive depositional environments seaward. (Stanley and Swift 1976, Fig. 4; published by permission of the authors and John Wiley and Sons.)

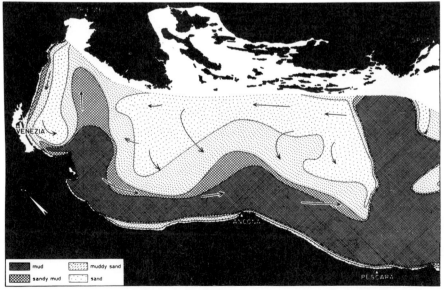

Figure 3.23 Distribution of sediment in the northern Adriatic Sea. (Van Straatten 1965, Fig. 60; published by permission of the author and Butterworths.)

Table 3.1. The Principal Processes Affecting Sedimentation in Outer Margin Environments[a]

Predominant controlling forces	Air/water interface	Water/sediment interface	Predominant transport mechanisms
Atmospheric and tidal	Wave agitation Wind-driven surface currents Eolian transport Glacial transport Tidal surface currents	Storm stirring on outer shelf Pelagic settling Tidal bottom currents Tidal canyon currents	Suspension, with subordinate traction
		Internal waves	
Thermohaline	Geostrophic surface currents	Geostrophic bottom currents	Suspension = traction
Gradient and gravity		Sliding and slumping Creep Debris flow Grain flow Turbidity flow	Mass movement, traction, and suspension all may be important
Biologic activity		Biogenic accumulation (tests, fecal pellets, etc.) Bioturbation and internal disturbance Surface erosion	Mainly *in situ* processes but some suspension; also traction or mass movement on steep slopes

[a] *From Stanley and Swift (1976, Chapter 17, Table 1). Published by permission of the authors and John Wiley and Sons.*

Aoyagi, K., N. Kobayashi, and T. Kazama. 1976. Clay mineral facies in argillaceous rocks of Japan and their sedimentary petrological meanings. *In:* S.W. Bailey, Ed., Proc. Internatl. Clay Conf., Mexico City, Mexico, 16–23 July 1975. Wilmette Publ. Ltd., Wilmette, Illinois, p. 101–110.

   Clay mineral associations of argillaceous rocks are divided into five major facies and four subfacies, the geologic controls being source rocks, depositional environment, diagenesis, metamorphism, and weathering. Postdepositional alteration is believed to be largely controlled by overburden pressure and geothermal temperature.

Byrne, P.J., and R.N. Farvolden. 1959. The Clay Mineralogy and Chemistry of the Bearpaw Formation of Southern Alberta. Alberta Res. Council Bull. 4, Calgary, Alberta, 44 pp.

   In this early study three vertical profiles of a widespread argillaceous formation show little environmental control on clay mineral diagenesis, although the authors suggest that abundant volcanic ash may mask some diagenesis. Chemical as well as clay mineral analyses.

• Carroll, D. 1970. Clay Minerals: A Guide to Their X-Ray Identification. Geol. Soc. Amer. Spec. Paper 126, 80 pp.

   An up-to-date reference on identification and quantitative estimates of abundance. Table 2 (not shown) lists the most common nonclay minerals in the less than 2-μm fraction.

Chen, Pei-Yuan. 1977. Table of Key Lines in X-Ray Powder Diffraction Patterns of Minerals in Clays and Associated Rocks. Indiana Geol. Survey Occas. Paper 21, 67 pp.

   A rapid guide to the X-ray identification of about 240 mineral species commonly found in shales. Twenty-one appendices.

• Cubitt, J.M. 1975. A regression technique for the analysis of shales by X-ray diffractometry: Jour. Sed. Petrol., v. 45, p. 546–553.

   One of the most recent of a series of papers on "automating" the mineralogic analysis of a shale. Thirty-one references.

Edzwald, J.K., and C.R. O'Melia. 1975. Clay distributions in Recent estuarine sediments: Clays Clay Minerals, v. 23, p. 39–44.

   Can clay minerals differentially flocculate in a salinity gradient? Field studies showed kaolinite landward of illite as predicted by laboratory work. Possible paleosalinity indicator if used in a careful mapping study. Compare with Parham's (1966) review of lateral variations observed in ancient sediments and Gibbs' (1977) work on the Amazon. See Fig. 3.24

• Fellows, P.M. and D.A. Spears. 1978. The determination of feldspars in mudrocks using an X-ray powder diffraction method: Clays Clay Minerals, v. 26, p. 231–236.

   There are almost no data on the amount and distribution of feldspar in fine-

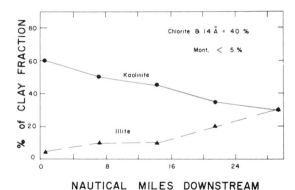

Figure 3.24 Clay mineral composition of Pamlico sediments downstream from the railroad bridge at Washington, N.C. (Edzwald and O'Melia 1975, Fig. 6; published by permission of the authors and The Clay Mineral Society.)

grained rocks. Using this method, quartz, albite, and microcline can be measured simultaneously.

Gibbs, R.J. 1977. Clay mineral segregation in the marine environment: Jour. Sed. Petrol. v. 47, p. 237–243.

Reports an increase in smectite with a parallel decrease in kaolinite and clay mica downcurrent along the Guiana shelf from the mouth of the Amazon. Gibbs attributes this change not to differential flocculation, but to the smaller size of the smectite particles. (Compare Edzwald and O'Melia, 1975.)

Gieseking, J.E., Ed. 1975. Soil Components, v. 2, Inorganic Components. Springer-Verlag, New York, 684 pp.

Seventeen technical chapters, including not only the clays, but feldspars, allophane, quartz and its relatives, water, bioliths, heavy minerals, and the use of thermal analyses and infrared spectroscopy. Many, many references.

Griffin, G.M., and B.S. Parrott. 1964. Development of clay mineral zones during deltaic migration: Amer. Assoc. Petrol. Geol. Bull., v. 48, p. 57–69.

Clay mineralogy may show transgressive–regressive cycles in a basin that is receiving fine-grained sediment from a large stream and from local or longshore sources of different mineralogy, as shown by the Mississippi Delta.

Griffin, J.J., H. Windom, and E.D. Goldberg. 1968. The distribution of clay minerals in the world ocean: Deep-Sea Res., v. 15, p. 433–459.

One of the most definitive papers ever written on the origin of clay minerals. This paper is based on geology's oldest technique—systematic mapping—which shows that climate plays a significant role in clay mineral composition at low latitudes, but not at high latitudes. This and other evidence indicates that the vast majority of clays in the recent sediments of the world ocean are detrital.

Grimshaw, R.W. 1971. The Chemistry and Physics of Clays. 4th ed. John Wiley and Sons, New York, 1032 pp.

> In addition to the ceramic properties of clays, this reference book covers the geology, crystal structure, identification methods, and chemistry of clays and other ceramic raw materials. A good source of fundamental data on the basic structures of the silicates and the nature of water on mineral surfaces. Contains 264 illustrations and numerous tables.

Hathaway, J.C. 1972. Regional clay mineral facies in estuaries and continental margin of the United States east coast. *In:* B.W. Nelson, Ed., Environmental Framework of Coastal Plain Estuaries. Geol. Soc. Amer. Mem. 133, p. 293–316.

> Eleven maps of diverse minerals in Holocene sediments, plus two inferred paleocurrent systems and one map of bottom drift directions. More than 400 samples from Key West to the Gulf of Maine.

Hayes, J.B. 1970. Polytypism of chlorite in sedimentary rocks: Clays Clay Minerals, v. 18, p. 285–306.

> Review of chlorite crystallography that suggests a stable IIb polymorph, which would be considered detrital if found in a shale, and a series of authigenic Ib polymorphs. Distinguishing these two types would be very valuable in studies of diagenesis, but the presence of illite interferes with the analysis and so makes the technique unusable for most older shales. See Fig. 3.25.

Jacobs, M.B., and J.D. Hays. 1972 Paleo-climatic events indicated by mineralogical changes in deep-sea sediments: Jour. Sed. Petrol., v. 42, p. 889–898.

> An attempt to relate clay mineralogy to climate in a modern, extensive, fine-grained deposit. Increased glacial activity in the Plio-Pleistocene is correlated with a dilution of smectite by illite/chlorite, probably caused by increased continental erosion. Such a change could be destroyed by later diagenesis, however.

Jeans, C.V. 1978. The origin of the Triassic clay assemblages of Europe with special reference to the Keuper Marl and Rhaetic of parts of England: Proc. Philosophical Trans. Royal Soc., Ser. A, v. 289, p. 549–636.

> Very comprehensive and exceptionally well done regional study of the clay mineralogy (140 samples) of a famous stratigraphic unit. Two major clay mineral assemblages are recognized: a detrital assemblage of mica and minor chlorite and a more limited neoformed assemblage of magnesium-rich clays (sepiolite, palygorskite, Mg-chlorite, smectite, corrensite, etc.). The clay mineralogy is very carefully intergrated with sedimentology and facies of which mudstone is a major part. Many informative illustrations. Outstanding and deserving to be often emulated in years to come.

Kabata-Pendias, A. 1967. Charakterystyka Geochemiczna Utworow Triasu z Rejonu Polski Polnocno-Zachodniej: Kwartalnik Geologiczny, Instytut Geologiczny, Warsaw, Poland, v. 11, p. 599–617.

> A clay mineral and trace element study of 368 bore hole samples from Buntsandstein equivalents. Primarily uses the hydromica–kaolinite ratio to infer dep-

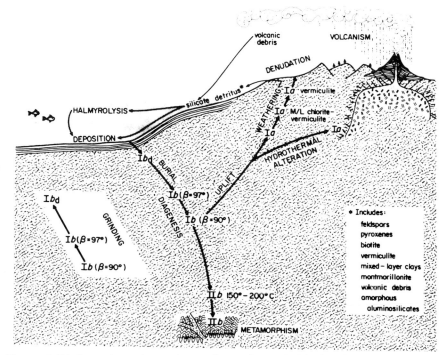

Figure 3.25 Genesis and transformations of type-I chlorites in relation to the rock cycle. (Hayes 1970, Fig. 13; published by permission of the author and The Clay Mineral Society.)

ositional environments. This ratio ranges from 0.2 (brackish to fresh water) to 12.0 (normal marine to saline). Polish with Russian and English summaries.

Keller, W.D. 1970. Environmental aspects of clay minerals: Jour. Sed. Petrol., v. 40, p. 788–813.

A useful, short review of a long-discussed topic by a well-known clay mineralogist. Rich in both philosophy and references. Suggests that mapping clay mineral composition laterally is a much better means of identifying the depositional environment than are "grab" samples. Good coverage of the English-language literature. Nine tables.

Main, M.S., P.F. Kerr, and P-K. Hamilton. 1950. Occurrence and Microscopic Examination of Reference Clay Mineral Specimens. American Petroleum Institute Project 49, Clay Mineral Standards, Prelim. Rept. 5, p. 15–57.

Divided into two parts with the second on microscopic examination, stressing the immersion method for identification of the optical indices of the clay minerals. Plates 1, 2, and 3 (not shown) contain three drawings of photomicrographs that show kaolinite mostly in terrigenous sediments.

McBride, E.F. 1974. Significance of color in red, green, purple, olive, brown, and gray beds of Difunta Group, northeastern Mexico: Jour. Sed. Petrol., v. 44, p. 760–773.

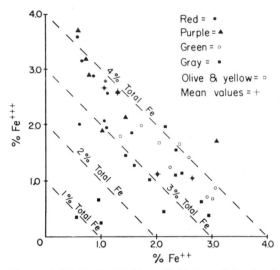

Figure 3.26 Plot of Fe^{2+} vs. Fe^{3+} from Difunta claystone with different colors. (McBride 1974, Fig. 11; published by permission of the author and the Society of Economic Paleontologists and Mineralogists.)

Field study of Cretaceous-Paleocene red, green, and purple mudstones and siltstones, plus their clay mineralogy and some chemical analyses. See Fig. 3.26 and Table 3.2.

● Moore, D.M. 1978. A sample of the Purington Shale prepared as a geochemical standard. Jour. Sed. Petrol., v. 48, p. 995–998.

Fourteen oxides, percent organic matter and 25 trace elements determined in a well known shale of the Illinois Basin.

Morton, R.A. 1972. Clay mineralogy of Holocene and Pleistocene sediments, Guadalupe Delta of Texas: Jour. Sed. Petrol., v. 42, p. 85–88.

Clay mineralogy in this near-shore sequence is controlled by source area rather than by diagenesis.

Naidu, A.S., D.C. Burrell, and D.W. Hood. 1971. Clay mineral composition and geologic significance of some Beaufort Sea sediments: Jour. Sed. Petrol., v. 41, p. 691–694.

Kaolinite is a common constituent of these high-latitude sediments, showing that it is not confined to areas of tropical weathering.

Neiheisel, J. 1972. Techniques for use of organic and amorphous materials in source investigations of estuary sediments. In: B.W. Nelson, Ed., Environmental Framework of Coastal Plain Estuaries. Geol. Soc. Amer. Mem. 133, p. 359–381.

An unusual provenance paper because it is concerned with the fine fraction and its organics plus amorphous materials. Useful model and technique paper for environmental sedimentology?

Table 3.2 Chemical analysis of red and green claystone samples, Cerro Huerta Formation, Difunta Group[a]

	Weight %	
	Red Sample 15	Green Sample 28
SiO_2	52.2	50.0
Al_2O_3	14.3	13.6
Fe_2O_3	4.01	1.21
FeO	1.41	3.96
CaO	8.54	9.54
MgO	2.19	3.20
Na_2O	1.08	0.88
K_2O	3.00	2.46
H_2O+	4.91	5.46
H_2O-	0.53	0.60
TiO_2	0.69	0.61
P_2O_5	0.18	0.19
MnO_2	0.03	0.05
CO_2	6.63	7.55

[a]From McBride (1974, Table 3). Published by permission of the author and the Society of Economic Paleontologists and Mineralogists.

Parham, W.E. 1966. Lateral variations of clay mineral assemblages in modern and ancient sediments. *In:* K. Gekker and A. Weiss, Eds., Proc. Internatl. Clay Conf., v. 1. Pergammon Press, London, p. 135–145.

An unusual and interesting paper that compiles much literature about a neglected subject. Author's illustrations effectively tell his story—that kaolinite is most abundant near the shoreline. Many more such studies are needed for mudstones and shales. See Fig. 3.27.

Pelzer, E.E. 1966. Mineralogy, geochemistry and stratigraphy of the Besa River Shale, British Columbia: Canadian Petrol. Geol. Bull., v. 14, p. 273–321.

A thick, 320–2400 m, shelf-to-basin section of a Devonian–Mississippian black shale is examined with the X-ray and resulting mineralogic variations are related to provenance. Maps and cross sections also used to infer such an origin. See also Conant and Swanson (1961).

Perrin, R.M.S. 1971. The Clay Mineralogy of British Sediments. Mineralogical Soc. (Clay Minerals Group), London, 247 pp.

A unique book that gives detailed analyses by stratigraphic unit with emphasis upon lithology as well. Useful summaries of each system are given. A model study that should be more widely followed.

Porrenga, D.H. 1967. Glauconite and chamosite as depth indicators in the marine environment: Mar. Geol., v. 5, p. 495–501.

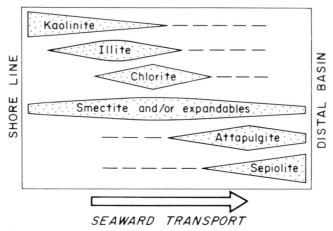

SEAWARD TRANSPORT

Figure 3.27 Schematic variation of clay minerals from shoreline to deep, distal basin. (Redrawn from Parham 1966, Fig. 6.)

Chamosite forms in shallow tropical waters (depths less than 60 m), and glauconite forms in deeper waters (30–2000 m) in temperate and tropical areas. Possible temperature/depth indicators? See Fig. 3.28.

Potter, P.E., D. Heling, N.F. Shimp, and W. Van Wie. 1975. Clay mineralogy of mod-

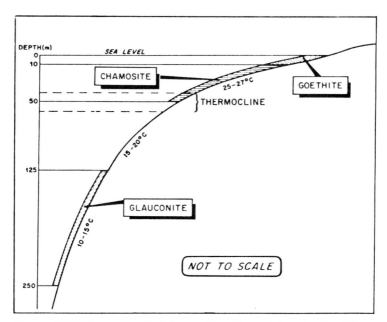

Figure 3.28 Generalized picture of the Niger delta, depth of the sea-floor, occurrence of goethite, chamosite and glauconite, and bottom-water temperature as measured in January and February 1959. (Porrenga 1967, Fig. 2; published by permission of the author and *Marine Geology*.)

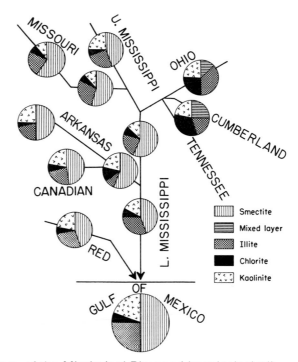

Figure 3.29 Clay Mineralogy of the Mississippi River and its principal tributaries reveals a western source of smectite and an eastern source of illite and chlorite and shows that most of the muds deposited by the Mississippi in the Gulf of Mexico come from its western tributaries. (Potter and others 1975, Fig. 13.)

ern alluvial muds of the Mississippi River Basin: Bull. Centre Rech. Pau-SNPA, v. 9, p. 353–389.

A provenance study, based on 177 samples of a 3 220 900-km² basin, finds only the two major mineral associations shown in Fig. 3.29. Authors conclude that in temperate and cold climate river basins, the composition of alluvial muds largely reflects source rocks.

Quakernaat, J. 1968. X-Ray Analysis of Clay Minerals in Some Recent Fluviatile Sediments along the coasts of Central Italy. Univ. Amsterdam, Physical Geography Laboratory, Publ. No. 12, 100 pp.

Systematic mapping of clay minerals in many small rivers shows mineralogic variation to depend primarily on source rocks. Compare with Chamley and Picard (1970).

Radan, S. 1975. Some data on the clay mineralogy and sedimentation in the abyssal zone of the Black Sea. *In:* Prima Conferinta Nationala Pentru Argile Bucurrsti, Nov. 1973. Institutul de Geologic si Geofisica, Ser. I (Mineralogie–Petrogrofie), No. 13, p. 113–125.

Grain size, organic carbon, clay mineralogy, illite crystallinity, etc., of 11.54 m
of core taken in the deepest part of the Black Sea.

Rateev, M.A., Z.N. Gorbumova, A.P. Lisitsyn, and G.L. Nosov. 1969. The distribution of clay minerals in the oceans: Sedimentology, v. 13, p. 21–43.

Primarily a study of the fine fraction (<0.001 mm) in the Indian and Pacific
Oceans shows kaolinite, gibbsite, and smectite to be maximal in the humid
tropics, and chlorite and illite most abundant in moderate to high latitudes. See
also Griffin and others (1968).

• Shaw, D.B., and C.E. Weaver. 1965. The mineralogical composition of shales: Jour.
Sed. Petrol., v. 35, p. 213–222.

Three hundred samples of Phanerozoic shales from the United States and
Canada were systematically and quantitatively studied by X-ray for quartz, feldspar, carbonate, and clay minerals yielding average values of 33.6, 3.6, 2.7, and
64.1%, respectively.

Sheppard, R.A., and A.J. Gude, 3rd. 1973. Zeolites and Associated Authigenic Sili-
cate Minerals in Tuffaceous Rocks of the Big Sandy Formation, Mohave County,
Arizona. U.S. Geol. Survey Prof. Paper 830, 36 pp.

Interbedded mudstones and tuffs show three diagenetic facies (nonanalcime
zeolites, analcime, and K-feldspar) related apparently to salinity and pH in the
pore waters of the lake from which they were deposited.

Spears, D.A. 1970. A kaolinite mudstone (tonstein) in the British Coal Measures:
Jour. Sed. Petrol., v. 40, p. 386–394.

Tonsteins are very thin shales of great lateral extent composed of kaolinite.
Spears argues that this one was formed by alteration of volcanic ash, based on
the restricted heavy mineral suite. Sixty references. Other examples of tonsteins
are discussed by Skocek (1973).

Suchecki, R.K., E.A. Perry, and J.F. Hubert. 1977. Clay petrology of Cambro-Ordovician continental margin, Cow Head Klippe, western Newfoundland: Clays
Clay Minerals, v. 25, p. 163–170.

A developing island arc in the source area is detected by an evolution of clay
mineralogy from an illite–chlorite derived from preexisting sediments to
illite–expandable chlorite and finally corrensite–illite–smectite from volcanic
sources. Compare Björlykke (1974).

Sudo, T., and S. Shimoda, Eds. 1978. Clays and Clay Minerals of Japan (Developments in Sedimentology, v. 26). Elsevier, North Holland Publ. Co., Amsterdam-
Oxford-New York, 344 pp.

Topics covered are chiefly of a general interest and include a summary of clays
and clay minerals in Japan, weathering of volcanic ash, wall rock alteration,
allophane and imogolite plus chapters on kaolinite, smectites, chlorites, and interstratified minerals. Many references.

Thompson, G.R., and J. Hower. 1975. The mineralogy of glauconite: Clays Clay Minerals, v. 23, p. 289–300.

The best study to date of this important mineral. Because it has been thought to be authigenic, glauconite is one of the few clays the composition and mineralogy of which may be useful in deciphering conditions in the environment of deposition. It is apparently an iron-rich, mixed-layer clay closely analogous to aluminous illite–smectite. Still to be decided is whether glauconites are truly authigenic or result from the substitution of iron for aluminum in preexisting clays.

Thorez, J. 1975. Phyllosilicates and Clay Minerals: A Laboratory Handbook for Their X-Ray Diffraction Analysis. Editions G. Lelotte, Dison, Belgium, 579 pp.

A massive English and French reference for the identification of clay minerals. Many charts and keys, some of which are colored, plus many references; seven chapters. See Fig. 3.30.

van Houten, F.B. 1962. Cyclic sedimentation and the origin of analcime-rich Upper

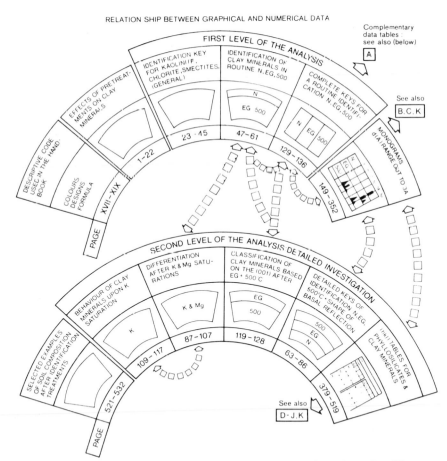

Figure 3.30 Flow diagram for comprehensive study of clay minerals. (Thorez 1975, Table of Contents; published by permission of the author and Editions G. Lelotte.)

Triassic Lockatong Formation, west-central New Jersey and adjacent Pennsylvania: Amer. Jour. Sci., v. 260, p. 561–576.

A thick lens of argillite of lacustrine origin with minor black shale and dark gray mudstone and some fine-grained sandstone has a cyclic succession and is rich in analcime. Some of the black shales have varve counts suggesting a 21 000-year precession cycle. See also van Houten (1964).

Van Olphen, H. and J. J. Fripiat, Eds. 1979. Data Handbook for Clay Materials and other Non-Metallic Minerals. Pergamon Press, Oxford, New York, Toronto, 346 pp.

Two parts provide those studying clays with sets of authoritative data describing the physical and chemical properties and mineralogical composition of the available reference materials for virtually all the possible ways of analyzing clays.

Weaver, C.E. 1963. Interpretive value of heavy minerals from bentonites: Jour. Sed. Petrol., v. 33, p. 343–349.

A literature review that examined 51 bentonites for heavy minerals: primary bentonites commonly have biotite, idiomorphic zircon, apatite and titanite, with some hornblende, augite and hypersthene, whereas secondary redeposited bentonites commonly have tourmaline, garnet, and muscovite. Notes that, in general, Tertiary and Recent bentonites have a wider range of heavy minerals than older ones. See also Blatt and Sutherland (1969) for discussion of intrastratal solution of heavy minerals in shales.

Weaver, C.E. and L.D. Pollard. 1973. The Chemistry of Clay Minerals. (Developments in Sedimentology, No. 15), Elsevier Scientific Publ. Co., Amsterdam, 205 pp.

Review of reported chemical analyses of each clay mineral type with a discussion of the relationship between structure and chemistry. About 500 references.

Wise, S.W., and F.M. Weaver. 1973. Origin of cristobalite-rich Tertiary sediments in the Atlantic and Gulf coastal plain: Trans. Gulf Coast Assoc. Geol. Socs., v. 23, p. 305–323.

Siliceous claystones are common in the Tertiary of the Gulf Coast and Atlantic coastal plains. This paper suggests that they were deposited in open marine water and, based on SEM work, that their high silica content results from the dissolution and reprecipitation of siliceous microorganisms.

• Yaalon, D.H. 1962. Mineral composition of average shale: Clay Minerals Bull., v. 5, p. 31–36.

Uses a slight modification of the method of Imbrie and Poldevaart (1959) to calculate normative mineral compositions from chemical analyses with explicit assignment of oxides into sedimentary minerals. Two comparative tables plus discussion of results. Believes the average shale to contain about 60% clay minerals. Useful technique paper. Compare with Shaw and Weaver (1965).

Scott W. Starratt
Dept. of Paleontology
U. C. Berkeley
Berkeley, Ca. 94720

GEOCHEMISTRY

The geochemistry of shales is an area of study that has seen much effort. Inorganic geochemical studies have mostly been of trace elements, and so far these have dealt primarily with the assignment of the various elements to phases within the rock, rather than with provenance or depositional history. An exception is the attempt, largely unsuccessful, to relate trace elements to paleosalinity. Major element geochemistry is less often seen and probably would yield interesting rusults if combined with careful mineralogy and stratigraphy.

Organic geochemical studies have been mostly devoted to understanding fossil fuel development or to searches for biochemical fossils, but a few papers have been published describing the stratigraphic distribution of organic compounds in shales. Because the organic content of shales is relatively high, the use of organic compounds in solving geologic problems appears to be a promising area for the innovative; in short what is needed is a *stratigraphically oriented geochemistry*—organic as well as inorganic.

Albrecht, P., M. Vandenbroucke, and M. Mandengué. 1976. Geochemical studies on the organic matter from the Douala Basin (Cameroon). I. Evolution of the extractable organic matter and the formation of petroleum: Geochim. Cosmochim. Acta, v. 40, p. 791–799.

> Good example of the successful work now being done on the diagenesis of organic matter using a coupled gas chromatograph–mass spectrometer.

• Baker, D.R. 1972. Organic geochemistry and geological interpretations: Jour. Geol. Education, v. 20, p. 221–234.

> One of the best summary papers on organic matter in shales, because it emphasizes the use of organic geochemical data in solving *geologic* problems.

• Björlykke, K. 1974, Geochemical and mineralogical influence of Ordovician island arcs on epicontinental clastic sedimentation. A study of Lower Paleozoic sedimentation in the Oslo region, Norway: Sedimentology, v. 21, p. 251–272.

> Provenance study of a 1000-m thick sequence based on chemical and clay mineral analyses. A chemical maturity index, defined as $(Al_2O_3 + K_2O)/(MgO + Na_2O)$, and trace elements are used to help determine provenance.

Bowie, S.H.U., J. Dawson, M.J. Gallagher, and D. Ostle. 1966. Potassium-rich sediments in the Cambrian of northwest Scotland: Trans. Inst. Mining Metallurgy, v. 75, B125–B145, and Discussion in v. 76, 1966, p. B60–B69.

> Lower Cambrian shales of Scotland, referred to as the Fucoid Beds, have very high K_2O contents, sometimes 11%. The paper and subsequent discussion center around possible economic uses of the shale as a long-term fertilizer, and its origin, various workers arguing for a syngenetic deposition of the K_2O or a later introduction by potassium-rich fluids. The references cited show that K-rich sedi-

mentary rocks are exceptionally abundant in the Cambro-Ordovician and so may reflect some nonuniform chemical condition.

Cazes, P., and Y. Reyre. 1976. La fossilisation du kérogène en milieu argilo-carbonáte; Etudé statistique des ses liaisons avec les propriétés lithologiques et pétrologiques dans l'Oxfordien du Bassin de Paris (partie orientale): Bull. Bur. Rech. Geol. Minieres, France, Ser. 2, Sec. 4, p. 85−102.

> Two hundred and seventy-seven samples from the calcareous clays of the Oxfordian of the Paris Basin were studied for their kerogen characteristics and related to the lithologic character of their host sediments, with emphasis on their petrology, sedimentary structures, and geologic occurrence. Factor analysis was used to help sort out relative importance of these and other diverse controlling factors.

● Cody, R.D. 1971. Adsorption and reliability of trace elements as environment indicators of shales: Jour. Sed. Petrol., v. 41, p. 461−471.

> This review paper notes that trace element concentration in shales depends on grain size and clay type, as well as on pH and salinity of original depositional environment. Hence, only broad, rather than minor, differences in paleosalinity can be ascertained.

● Degens, E.T., E.G. Williams, and M.L. Keith. 1957. Environmental studies of Carboniferous sediments, Part I: Geochemical criteria for differentiating marine and fresh-water shales: Amer. Assoc. Petrol. Geol. Bull., v. 41, p. 2427−2455.

> One of the first papers to study the basin-wide distribution of trace element assemblages of shale units. Compares trace element assemblages to stratigraphic and paleontologic interpretations of depositional environments of Pennsylvanian cyclothems in the western Appalachian Basin.

Englund, J-O., and P. Jorgensen. 1973. A chemical classification system for argillaceous sediments and factors affecting their composition: Geol. fören Stockholm Förh., v. 95, p. 87−97.

> Plots of $K_2O + Na_2O + CaO, MgO + FeO$, and Al_2O_3 on a triangle for a wide range of argillaceous sediments. Factors influencing geochemical parameters include bedrock composition, weathering, and grain size.

Fritz, B. 1975. Étude Thermodynamique et Simulation des Réactions entre Minéraux et Solution. Application a la Géochimie des Alterations et des eaux Continentales: (Bur. Rech. Geol. Minieres-Institut de Geologie, Strasbourg, Memoire Sciences Geologiques, No. 41, 152 pp.

> Geologists are helped to understand the geochemistry of alteration by continental waters by using thermodynamics together with a mathematical simulation of the reactions between minerals and solutions found in nature.

Hudson, J.D. 1978. Concretions, isotopes, and the diagenetic history of the Oxford Clay (Jurassic) of central England: Sedimentology, v. 25, p. 339−370.

> Carbon and oxygen isotopes are used to decipher the diagentic history of this shale. The earliest found mineral assemblage was calcite + pyrite. Those

concretions that formed before compaction were subsequently breciated, and some were infilled with barite. The final event was precipitation of a ferroan calcite that may be related to an influx of fresh ground water.

Jux, U., and U. Manze. 1974. Milieu-Indikationen im Devon des Bergischen Landes mittels Kohlenstoff-Isotopen. Neues Jahrb. Geol. Paläeontol., Monatsh., n. 6, p. 353–373.

Concretions in the Devonian black shales of Germany have thicker laminations than in the associated shales and studies of the stable isotopes of C^{12}/C^{13} indicate a shift from light to heavier isotopic ratios—a gradual deepening of the sea?

Katada, M., H. Isomi, E. Omori, and T. Yamada. 1963. Chemical composition of Paleozoic rocks from northern Kiso District and of Toyoma Clay Slates in Kitakami Mountainland: I. Chemical composition of pelitic rocks: Japanese Assoc. Mineral. Petrol. Econ. Geol., v. 49, p. 85–100.

Nineteen chemical analyses, some of which are metamorphic argillites.

Kepferle, R.C. 1959. Uranium in Sharon Springs Member of Pierre Shale, South Dakota and Northeastern Nebraska. U.S. Geol. Survey Bull. 1046-R, p. 577–604.

Radioactivity in this thin, persistent black shale was mapped regionally and found to increase away from the source area, presumably in response to increasing amounts of organic matter.

Kimble, B.J., J.R. Maxwell, R.P. Philip, G. Eglinton, P. Albrecht, A. Ensminger, P. Arpino and G. Ourisson. 1974. Tri-and tetraterpenoid hydrocarbons in the Messel Oil Shale: Geochim. Cosmochim. Acta, v. 38, p. 1165–1181.

An example of the use of organic chemical techniques to identify the types of organisms that contributed organic matter to the sediment. Comparisons are made with similar work on the Green River Shale.

Manheim, F.T., J.C. Hathaway, F.J. Flanagan, and J.D. Fletcher. 1976. Marine mud, MAG-1, from the Gulf of Maine. In: F.J. Flanagan, Ed., Descriptions and Analyses of Eight New USGS Rock Standards. U.S. Geological Survey Prof. Paper 840, p. 25–28.

Thin section and X-ray analyses plus major oxides and trace elements. See Schultz, et al. for comparable data on a shale.

Prashnowsky, A.A. 1971 Biogeochemische Untersuchungen an Tonsteinen des Ruhrkarbons: Geol. Rundschau, v. 60, p. 744–812.

A very comprehensive study of the organic geochemistry of 18 "coal tonsteins," tonsteins being clay partings associated with coals. Reports on their amino acids, carbohydrates, lipids, and humic acids, with much tabled and graphic data, which is related to stratigraphy and compared to that of associated coals.

Roaldset, E. 1972. Mineralogy and geochemistry of Quaternary clays in the Numedal area, southern Norway: Norsk Geol. Tidsskrift, v. 52, p. 335–369.

Complete clay mineralogy and chemical analysis with emphasis upon geochemical variations introduced by weathering. Uses Vogt's maturity ratio. Much tabled data and good figures with 66 references. Is there a North American analog to this study?

•Schultz, L.G., H.A. Tourtelot, and F.J. Flanagan. 1976. Cody Shale SCO-1, from Natrona County, Wyoming. *In:* F.J. Flanagan, Ed.: Descriptions and Analayses of Eight New USGS Rock Standards. U.S. Geological Survey Prof. Paper 840, p. 21–23.

Thin section and X-ray analysis plus major oxides and trace elements. See Manheim, *et al.* (1976) for comparable data on a modern mud.

•Shaw, D.M. 1956. Geochemistry of pelitic rocks. III: Major elements and general geochemistry: Geol. Soc. Amer. Bull., v. 67, p. 919–934.

One hundred and fifty-five analyses yielded a grand average. Also, no change in composition was found going from mud to shale to slate to schist and gneiss beyond loss of H_2O and CO_2. Comparison with the average igneous rock. Good source of data, but most of the reported analyses fail to distinguish exchangeable and nonexchangeable cations. See also Shaw's (1954) paper in Geol. Soc. Amer. Bull., v. 54, p. 1151–1182.

Swain, F.M., and N.A. Rogers. 1966. Stratigraphic distribution of carbohydrate residues in Middle Devonian Onondaga beds of Pennsylvania and western New York: Geochim. Cosmochim. Acta, v. 30, p. 497–509.

One of the few organic geochemical studies that attempts to map the distribution of a substance. In this case (Fig. 4, not shown), carbohydrates were found to be uniformly low (25–50 ppm) over most of Pennsylvania, but there is a trough of very high concentration (300–400 ppm) in the northwest corner of the state, apparently caused by locally deeper water.

• Tardy, Y. 1975. Element partition ratios in some sedimentary environments. I: Statistical treatments. II: Studies on North American black shales: Sci. Geol., v. 28, p. 59–95.

Excellent treatment of geochemistry of trace elements in shales by a leading expert. Particularly interesting is the use of factor and regression analysis to assign elements to various mineral phases in the rock. Analyses of 1000 samples from 20 different black shale units were used to illustrate these techniques. Much information has been accumulated in recent years on the geochemical behavior of trace elements, but there has been surprisingly little success in using this geochemistry to solve geologic problems involving shales.

Tardy, Y., and R.M. Garrels. 1974. A method of estimating the Gibbs energies of formation of layer silicates: Geochim. Cosmochim. Acta, v. 38, p. 1101–1116.

Provides a much needed way to calculate the free energies of minerals of variable composition, such as illites and smectites, without resorting to difficult thermochemical measurements on each composition.

Tourtelot, H.A. 1962. Preliminary Investigation of the Geologic Setting and Chemical

Composition of the Pierre Shale, Great Plains Region. U.S. Geol. Survey Prof. Paper 390, 74 pp.

Data from 22 samples are supplemented by an additional 45 and subjected to a very complete, methodologic study. Although there is a summary of stratigraphy, most effort is geochemical. Figure 1 (not shown) is interesting, as is the section entitled "Geochemical History." See also U.S. Geological Survey Professional Pappers 391-A, 391-B, 392-A, 392-B, 393-A, and 393-B for related papers on the Pierre Shale.

Tourtelot, E.B. 1970. Selected Annotated Bibliography of Minor-Element Content of Marine Black Shales and Related Sedimentary Rocks, 1930–1965. U.S. Geol. Survey Bull. 1293, 118 pp.

Three hundred and seventy-five references to worldwide occurrences of black shale—a good working base for information on metalliferous, organic-rich black shales. Usefully cross indexed.

Vine, J.D., and E.B. Tourtelot. 1970. Geochemistry of black shale deposits: A summary report: Econ. Geol., v. 65, p. 253–272.

Seven hundred and seventy-nine samples of North American black shales were analyzed to establish normal range of compositions and define metal-rich black shale. Some implications of importance of black shales in ore-forming processes are discussed.

PETROLOGY

The advent of radiography promises new life to the petrologic study of shale, primarily because it permits us to see the fine details of their texture and fabric. Other useful techniques include one of the oldest in geology—the study of thin sections—as well as the SEM, X-ray diffraction, and studies of the internal surface area of shales. Another instrument we hope will be more widely used is the electron microprobe, especially when used jointly with an SEM.

It seems to us that the above techniques and instruments will be most effective when they are closely integrated with stratigraphic studies. If this is done, the resulting *stratigraphic petrology* will have the best chance of solving most geologic and engineering problems.

Finally, we would like to raise the question of reconstructing paleocurrent patterns from the fabric of shales. Which of the above petrologic techniques can do it best?

Bates, T.F., and E.O. Strahl. 1957. Mineralogy, petrography, and radioactivity of representative samples of Chattanooga Shale: Geol. Soc. Amer. Bull., v. 68, p. 1305–1313.

Short, informative paper with thin-section descriptions and general mineralogy of the Devonian black shale.

Bitterli, P. 1963. Aspects of the genesis of bituminous rock sequences: Geol. Mijnbouw, v. 42, p. 183–201.

A landmark paper based on a stratigraphic–petrologic study of samples from
western Europe, ranging in age from Cambrian to Tertiary. Suggests that
brackish and limnic deposits tend to be better producers of bituminous rocks
than open marine sediments and that they tend to be associated with
paleogeographic turning points, such as epeirogenic and/or eustatic oscilla-
tions, and thus orogeneses. Good colored plate with 16 photomicrographs.

● Blatt, H., and D.J. Schultz. 1976. Size distribution of quartz in mudrocks:
Sedimentology, v. 23, p. 857–866.

Quartz is the dominant nonclay mineral of mudrocks, where it is present as a
poorly sorted fine to medium silt. Authors believe that most of the world's detrital crystalline silica occurs in mudrocks.

• Bohor, B.F., and R.E. Hughes. 1971. Scanning electron microscopy of clays and clay
minerals: Clays and Clay Minerals, v. 19, p. 49–54.

Twenty-eight half-page SEM photographs of clay minerals described in terms of
configuration, texture, and growth mechanics.

Burger, K. 1963. Kaolin-Übergangstonstein, das genetisch-fazielle, bilaterale Bin-
deglied zwischen Kaolin-Kohlentonstein and Kaolin-Pseudomorphosentonstein.
Geol. Mitt., v. 4, p. 115–153.

Magnificent thin-section photomicrographs and fine drawings of the petrography of a tonstein, clay beds associated with coals, in the Ruhr. Excellent
example of a thin-section study of a shale.

• Carozzi, A.V. 1960. Microscopic Sedimentary Petrology. John Wiley and Sons, New
York, 485 pp.

One of the few petrography books with much on argillaceous sediments (pp.
124–189). Many European references.

Delmas, M.R. L'etude de la diagenèse des sédiments carbonates par l'utilization de
plaques ultra-minces. Bull. Centre Rerch. Pau—SNPA, v. 8, p. 95–109.

For carbonates, ultrathin slides, 10 μm thick, show a variety of textures not
revealed in standard thin sections. Could not the same be true for shales?
Three plates, one of which is colored.

Diamond, S. 1970. Pore size distribution in clays: Clays Clay Minerals, v. 18, p.
7–23.

Discussion of mercury porosimetry of clays and clay soils. This is a potentially
very powerful method for characterizing fabric that has seldom been success-
fully applied to shales. See Fig. 3.31.

Droste, J.B., and C.J. Vitaliano. 1973. Tioga Bentonite (Middle Devonian) of Indiana:
Clays Clay Minerals, v. 21, p. 9–13.

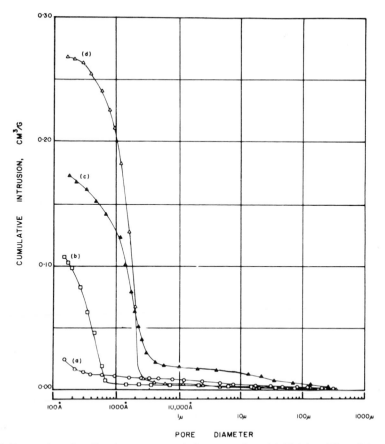

Figure 3.31 Pore-size distribution curves of clay samples: (a) Fithian illite; (b) Clay Spur montmorillonite; (c) Garfield nontronite; (d) Macon kaolinite. (Diamond 1970, Fig. 4; published by permission of the author and The Clay Mineral Society.)

Thin-section petrology and clay mineralogy support a volcanic origin for shale partings in a carbonate sequence. This bentonite layer can be traced on geophysical logs.

Folk, R.L. 1960. Petrography and origin of the Tuscarora, Rose Hill, and Keefer Formations, Lower and Middle Silurian of eastern West Virginia: Jour. Sed. Petrol., v. 30, p. 1–59.

This classic paper by a famous petrographer has some excellent thin-section descriptions of Silurian shales, descriptions that can be usefully copied. See Fig. 3.32.

Folk, R.L. 1962. Petrography and origin of the Silurian Rochester and McKenzie Shales, Morgan County, West Virginia: Jour. Sed. Petrol., v. 32, p. 539–578.

Useful as a possible model for meaningful thin-section descriptions of shale, al-

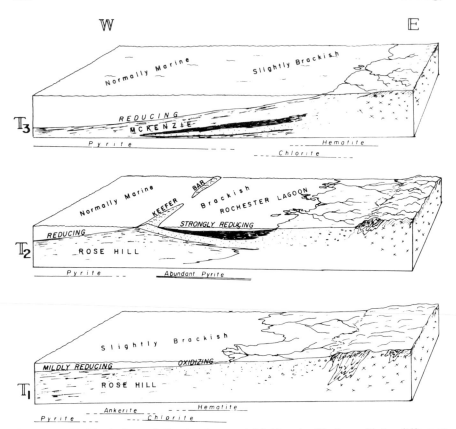

Figure 3.32 Evolution of Rochester and McKenzie Shales. Note different ranges of iron-rich minerals. (Folk, 1960, Fig. 12; published by permission of the author and the Society of Economic Paleontologists and Mineralogists.)

though most emphasis (as it should be?) is upon the interbedded thin limestones, siltstones, and sandstones.

Fraser, G.S., and A.T. James. 1969. Radiographic Exposure Guides for Mud, Sandstone, Limestone, and Shale. Illinois Geol. Survey Circ. 443, 20 pp.

Many graphs and clear text make this short article very useful for those who wish to see the internal structure of shales using a radiograph with a nondestructive preparation technique. Seventeen references.

• Gilliot, J.E. 1969. Study of the fabric of fine-grained sediments with the scanning electron microscope: Jour. Sed. Petrol., v. 39, p. 90−105.

An illustration of the use of SEM techniques in fine-grained sediments. Noteworthy in that argillaceous limestones are included.

Gilliot, J.E. 1970. Fabric of Leda Clay investigated by optical, electron−optical and X-ray diffraction methods: Eng. Geol., v. 4, p. 133−153.

An initial study of five samples from a Pleistocene clay relates clay mineral orientation to paleosalinity; i.e., clays with higher paleosalinites have poorer fabric orientation (because of flocculation) than do those with low paleosalinities.

• Gipson, M. Jr. 1965. Application of the electron microscope to the study of particle orientation and fissility in shale: Jour. Sed. Petrol., v. 35, p. 408–414.

The electron and petrographic microscope both show that in a Pennsylvanian black shale from Iowa, elongate particles, aggregates, and lenses show a strong preferred orientation in the bedding plane which is being stronger where the shale is fissile.

Gipson, M. Jr. 1966. A study of the relations of depth, porosity and clay mineral orientation in Pennsylvanian shales: Jour. Sed. Petrol., v. 36, p. 888–903.

In western Kentucky 26 samples were studied from a 1500-ft section of core and fairly complex interrelationships between depth, porosity, and clay mineral orientation were revealed using a simple correlation matrix. Probably most significant is that porosity depends more on the environment of deposition and fissility than on depth of burial alone.

Heling, D. 1969. Relationships between initial porosity of Tertiary argillaceous sediments and paleosalinity in the Rheintalgraben: Jour. Sed. Petrol., v. 39, p. 246–254.

Inferred initial porosity of shales (measured porosity normalized to a constant grain size and overburden thickness) was found to correlate well with paleosalinity, the nonmarine shales having much lower initial porosity. This effect is attributed to the unflocculated state of the nonmarine clays when deposited. Interesting application of a physical measurement to a geologic problem.

Heling, D. 1970. Micro-fabrics of shales and their rearrangement by compaction: Sedimentology, v. 15, p. 247–260.

The best paper to date on shale fabric. Grain-size distribution, total porosity, pore-size distribution (by mercury porosimetry), and specific surface area (by nitrogen gas adsorption) were measured. Down to 1000 m, these properties are controlled only by compaction, but below this zone, recrystallization effects become important.

Keller, W.D. 1976. Scan electron micrographs of kaolins collected from diverse environments of origin—I, II and III: Clays Clay Minerals, v. 24, p. 107–113, 114–117, and 262–264.

Sixty-eight SEM photographs of diverse clays, plus a short text by a famous clay mineralogist, one who has long emphasized the geology of clay minerals.

Loreau, J.-P. 1970. Contribution á l'étude des calcarénites hétérogènes par l'emploi simultané de la microscope photonique et de la microscope électronique á balayage. Problème particular de la micritisation: Jour. Micros., v. 9, p. 727–734.

Superb photomicrographs, 19 in all, by the SEM of micritic textures of carbonate mud and of the micritization process of fossils. Scanning electron microscope photographs compared with those from the ordinary microscope. Unusual technique paper that could, it seems to us be profitably followed by students of shales. See also Delmas (1974).

Martin, R.T. 1966. Quantitative fabric of wet kaolinite: Clays Clay Minerals, v. 14, p. 271–287.

Describes a method of impregnating wet clays in order to prepare them for grinding a flat surface.

• Meade, R.H. 1961. X-Ray diffractometer method for measuring preferred orientation in clays. U.S. Geol. Survey Prof. Paper 424B, pp. 273–276.

A method of X-ray fabric analysis using polished disks, cut perpendicular and parallel to bedding.

Moreland, G., F. Ingram, and H.H. Banks Jr. 1972. Preparation of doubly polished thin sections. In: W.G. Melson, Ed., Mineral Science Investigations. Smithsonian Contributions to the Earth Sciences, 1969–1971, v. 9, p. 93–94.

Details of how to possibly improve your thin sections of shales as well as other sediments and rocks.

• Morgenstern, N.R., and J.S. Tchalenko. 1967. The optical determination of preferred orientation in clays and its application to the study of microstructures in consolidated kaolin—I. and II: Proc. Royal Soc. London, 300A, p. 218–250.

Although single clay particles can seldom be seen with an optical microscope, the preferred orientation of an aggregate of clay particles can be investigated by thin-section study of the birefringence of an aggregate and related to degree of preferred orientation. Key ideas: Clays, birefringence, and orientation.

O'Brien, N.R. 1968. Electron microscope study of black shale fabric: Naturwissenschaften, v. 55, p. 490–491.

Electron micrograph of the black shale above the Pennsylvanian Harrisburg (No. 5, not shown) Coal in Illinois shows lamination and stair-step arrangement of clay layers in fissile shale. No clumps of organic material observed and therefore possibly color results from absorbed organic molecules? Comparisons to X-ray methods.

• O'Brien, N.R. 1970. The fabric of shale—An electron-microscope study: Sedimentology, v. 15, p. 229–246.

Many fine microphotographs, SEM and transmission electron microscopy (TEM), of clay particles in shales. Degree of orientation correlated with fissility. Some discussion of causes of differences in orientation are given, but this needs to be confirmed by study of areal variations within a single shale unit.

Odom, I.E. 1967. Clay fabric and its relation to structural properties in mid-continent Pennsylvanian sediments: Jour. Sed. Petrol., v. 37, p. 610–623.

The degree of orientation of clay particles in shales, as measured by X-ray diffraction, increases with increasing organic matter and decreases with increasing carbonate content. See Fig. 3.33.

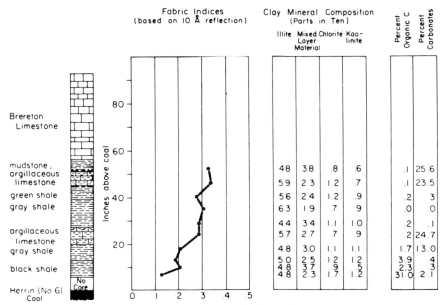

Figure 3.33 Clay fabric, clay mineral composition, and percent of organic carbon and carbonate minerals in sediments overlying the Harrisburg (No. 5) Coal, Franklin Co., Illinois, U.S.A. (Odom 1967, Fig. 2; published by permission of the author and the Society of Economic Paleontologists and Mineralogists.)

• Piper, D.J.W. 1972. Turbidite origin of some laminated mudstones: Geol. Mag., v. 109, p. 115–126.

> Many deep-water marine muds have thin, graded beds in which mud and silt alternate, with the slit being graded. Unusual plots of size versus distance *in* thin sections 15–25 mm long. Orientation of silt grains in plane of bedding also measured.

Pryor, W.A., and W.A. Van Wie. 1971. The "sawdust sand"; An Eocene sediment of floccule origin: Jour. Sed. Petrol., v. 41, p. 763–769.

> Fine-grained material was originally deposited in much larger aggregates, apparently produced by clay flocculation. Size analysis of disaggregated shale therefore may be very misleading.

Pusch, R. Microstructural features of Pre-Quaternary clays: Stockholm Contrib. Geol., v. 24, p. 1–24.

> An investigation of the geometric arrangement of clay size particles in incompletely lithifield shales in northern Europe, with special reference to clay mineralogy and stress history using a transmission electron microscope, SEM, and the light microscope, plus the X-ray. Data on size analysis, water content, and both particle orientation and particle-to-particle contacts. Salinity, clay mineralogy, stress history, saturating cations, and porosity are all believed to affect microstructure. Fabric and porosity are closely related to clay mineral com-

position and kaolinitic and illitic clays are observed to have tighter fabrics and lower porosity than the montmorillonitic clays. Pore size and porosity are much smaller in heavily loaded clays than in the less load influenced clays.

Scotford, D.M. 1965. Petrology of the Cincinnatian Series shales and environmental implications: Geol. Soc. Amer. Bull., v. 76, p. 193–222.

A number of chemical, mineralogic, and textural measurements of Ordovician shales show no vertical variations, but there is a horizontal separation along an east–west line for one unit, apparently produced by shallower water to the south. A rare study of a mixed shale–limestone sequence emphasizing areal variations.

, Siever, R., and M. Kastner. 1972. Shale petrology by electron microprobe; Pyrite–chlorite relations: Jour. Sed. Petrol., v. 42, p. 350–355.

Illustrates the use of the electron microprobe for study of diagenetic mineralogy in fine sands. A promising start that needs to be followed up by more workers.

Silverman, E.N., and T.F. Bates. 1960. X-Ray diffraction study of orientation in Chattanooga Shale: Amer. Mineral., v. 45, p. 60–68.

Degree of orientation of 001 planes of 10-Å illite measured in sections perpendicular to the bedding on 58 samples from the Chattanooga Shale. There was no correlation between orientation of illite and uranium content.

Skocek, V. 1973. Contribution to the problem of tonstein origin: Casopis Mineral. Geol., v. 18, p. 233–242.

A brief and clear discussion of the "tonstein problem" (tonsteins are thin, widespread kaolinitic clay beds often associated with the Carboniferous coals of Europe), with presentation of the possible origins of this enigmatic argillaceous rock. Some thin-section micrographs and a comprehensive bibliography.

Tchalenko, J.S., A.D. Burnett, and J.J. Hung. 1971. The correspondence between optical and X-ray measurements of particle orientation in clays: Clays Clay Minerals, v. 9, p. 47–70.

Study of experimental slurries and two natural clays suggests that there is a correspondence between optical and X-ray orientation of clay particles, the spatial arrangement of platy particles in a clay being defined by an *orientation ratio*. Both measurements can be made on the same thin section. However, authors suggest that the X-ray method is the better.

Urbain, P. 1951. Recherches Pétrographiques at Géochimiques sur Deux Séries de Roches Argileuses; 1°, Lias et Oolithique du Calvados; 2° Éocene et Oligocène de la région de Paris. Mémoirs de la Carte Géologique Détaillée de la France, Paris, Imprimerie Nationale, 278 pp.

A pioneer effort—with strong emphasis on petrology—to examine the argillaceous sediments of a Mesozoic and Tertiary sequence in France using the petrographic microscope, chemistry, size analysis, etc., plus 10 plates; many of the photomicrographs are of mudstones.

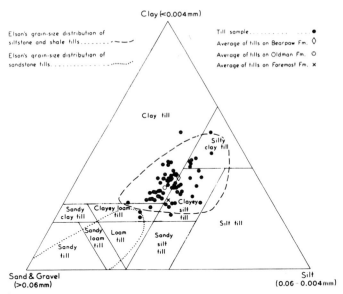

Figure 3.34 Textural classification of tills in the Foremost-Cypress Hills of western Canada. (Westgate 1968, Fig. 18.)

Weaver, C.E. 1958. Geologic interpretation of argillaceous sediments. Part II—Clay petrology of Upper Mississippian–Lower Pennsylvanian sediments of central United States: Amer. Assoc. Petrol. Geol. Bull., v. 42, p. 272–309.

Extensive study of clay mineral assemblages of related tectonic facies. Shows tectonism and provenance to be principal influence on shale mineralogic variations. Includes and compares petrologic analyses of associated sandstone and limestone facies, the result of several thousand clay mineral analyses. Basinal maps of clay mineral assemblages.

• Weaver, C.E. 1968. Electron microprobe study of kaolin: Clays Clay Minerals, v. 16, p. 187–189.

One of the rare examples of the use of this instrument on fine-grained material. Iron impurities were found to be evenly distributed over the kaolinite grains, instead of in discrete mineral impurities.

Westgate, J.A. 1968. Surficial geology of the Foremost–Cypress Hills area: Alberta Research Council Bull. 22, Calgary, Alberta, 121 pp.

A very comprehensive study that includes much on the petrography of clay-rich tills as shown by Fig. 3.34.

PALEONTOLOGY AND PALEOECOLOGY

We include here papers that have used fossils to help clarify the origin of the muds and shales in which they occur—their age, internal time stratigraphy, water depth, and paleoecology.

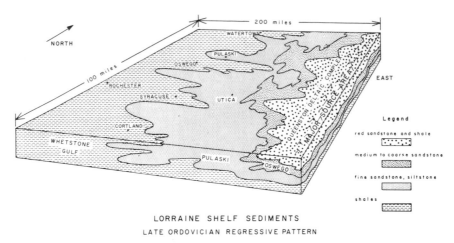

LORRAINE SHELF SEDIMENTS
LATE ORDOVICIAN REGRESSIVE PATTERN

Figure **3.35** Integrated environmental reconstruction of Late Ordovician communities in northcentral New York. (Bretsky 1970, Fig. 14.)

Some of these papers use specific groups—foraminifera, conodonts, or ammonites, for example—whereas others treat some aspects of the more general subject of organism-muddy substrate adaptation. Unfortunately, we could find no papers specially treating the general subject of how different organisms adapt to a muddy environment. In contrast, there is a rapidly growing awareness of the amount and significance of bioturbation in shales and, therefore, it was fairly easy to find relevant papers. Also included are several examples of how specific fossil groups are used to establish correlation in thick shale sections, especially conodonts in black shales.

The paper by Zangerl and Richardson (1963) and that by Hills and Levinson (1975), although very different in kind, are both good examples of how to integrate the fossil content of a shale with its other characteristics.

Anderson, H.V. 1961. Genesis and Paleontology of the Mississippi Mudlumps. Part II—Foraminifera of the Mudlumps, Lower Mississippi River Delta: Louisiana Geol. Survey Bull. 35, 208 pp.

Foraminifera of the mud lumps (cf. Morgan, 1961) were used to show that they must be derived from depths of at least 400 ft.

Bretsky, P. 1970. Late Ordovician Benthic Marine Communities in North-Central New York. New York State Museum and Sci. Ser. Bull. 414, 34 pp.

Paleoecologic study of a prograding delta–shelf sequence, including the Whetstone Gulf shale facies. Benthic communities, including trace fossils, are defined for each of the major facies. See Figs. 3.35 and 3.36.

Byers, C.W. 1974. Shale fissility: Relation to bioturbation: Sedimentology, v. 21, p. 479–484.

Suggests that there is an inverse relation between bioturbation and fissility

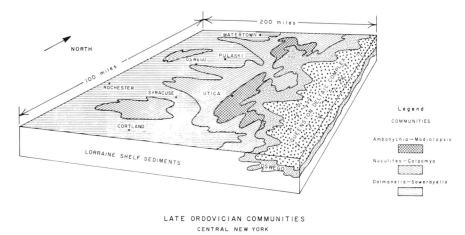

LATE ORDOVICIAN COMMUNITIES
CENTRAL NEW YORK

Figure 3.36 Reconstruction of Lorraine shelf sediments in New York. (Bretsky 1970, Fig. 3.)

(lamination) that depends on the oxygen content of the basin's water. Could not this idea be extended to color and carbon content as well?

Byers, C.W. 1977. Biofacies patterns in euxinic basins: A general model. Soc. Econ. Paleontol. Mineral. Spec. Publ. 25, p. 5–17.

Simple but elegant model relating oxygen in the overlying water to the amount and type of bioturbation in the sediment. This idea seems to have widespread application.

Dailey, D.H. Early Cretaceous Foraminifera from the Budden Canyon Formation, Northwestern Sacramento Valley, California: Univ. Calif. Publ. Geol. Sci., v. 106, 111 pp.

A paleoecologic study of an interesting sequence of dark shales with interbedded graywacke sandstones, conglomerates, and pebbly mudstones. Limestone concretions are common. Foraminifera indicate mostly outer neritic to bathyal depths for most of the sequence except the basal conglomerate, which was shallow water.

Frush, M.P., and D.L. Eicher. 1975. Cenomanian and Turonian foraminifera and paleonenvironments in the Big Bend region of Texas and Mexico. In: W.G.E. Caldwell, Ed., The Cretaceous System in the Western Interior of North America, Geol. Assoc. Canada Spec. Paper 13, p. 278–301.

Primarily of interest for its two different bathymetric models to explain thick marine shale facies of the interior plains that grade into thin carbonates to the east. Paleoecology of forams used to help identify correct model of regional mud deposition.

Gill, J.R., and W.A. Cobban. 1973. Stratigraphy and Geologic History of the Montana Group and Equivalent Rocks, Montana, Wyoming, and North and South Dakota. U.S. Geol. Survey Prof. Paper 776, 37 pp.

Figure 3.37 Inferred habitat of *Waagenoconcha abichi:* spat attached to idealized algae. Juveniles lie on soft mud substrate, and adults are in various stages of maturity and burial. (Grant 1966, Fig. 2; published by permission of the author and the Paleontological Society.)

Mostly paleontologic study of a group of sediments that contain several thick shales. Valuable because it uses careful biostratigraphic control to determine rates of deposition and of transgression–regression.

Grant, R.E. 1966. Spine arrangement and life habits of the productoid Brachiopod *Waagenoconcha:* Jour. Paleontol., v. 40, p. 1063–1069.

An example of the adaptation required of benthonic organisms when they live on a muddy bottom—in this case a foundation of many spines, as shown in Fig. 3.37.

Hills, L.V., and A.A. Levinson. 1975. Boron ccntent and paleoecologic interpretation of the Bearpaw and contiguous Upper Cretaceous strata in the Strathmore Well of Southern Alberta. *In:* W.G.E. Caldwell, Ed., The Cretaceous System in the Western Interior of North America, Geol. Assoc. Canada Spec. Paper 13, p. 411–415.

An integrated study of spores, Foraminifera and boron shows a high correlation with depositional environment. Would it not be useful to have also studied clay fabrics as did Heling (1969 and 1970)? See Fig. 3.38.

Howard, J.D., and R.W. Frey. 1975. Estuaries of the Georgia coast, U.S.A.: Sedimentology and biology: Senckenbergiana Maritima, v. 7, p. 1–305.

Seven papers on animal–sediment relations and sediment distribution in an area of modern mud accumulation. Much valuable information on use of biogenic

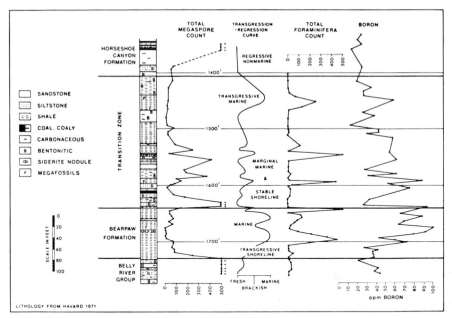

Figure 3.38 Megaspores, foraminifera and boron in the C.P.O.G. Strathmore E.V. 7-12-25-25 W4 well. (Hills and Levinson 1975, Fig. 1; published by permission of the authors and the Geological Association of Canada.)

and sedimentary structures to decipher environments, with numerous plates. Mostly devoted to description of the sands, but the papers by Howard and Frey, "Regional Animal–Sediment Characteristics," and by Mayou and Howard, "Animal–Sediment Relationships of a Salt Marsh Estuary—Doboy Sound," contain descriptions of the fine-grained sediments. Compare Pryor (1975).

Imbrie, J. 1955. Quantitative lithofacies and biofacies study of Florena Shale (Permian) of Kansas: Amer. Assoc. Petrol. Geol. Bull., v. 39, p. 649–670.

A very thin (usually less than 12 ft. thick) widespread, Permian marine shale is sampled at 18 points across eastern Kansas for lithology, macrobiofacies, and some sedimentary geochemistry (mostly strontium–calcium ratio) and used for an integrated paleoecologic interpretation. Possible model for systematic ecologic study of shales.

Izett, G.A., W.A. Cobban, and J.R. Gill. 1971. The Pierre Shale near Kremmling, Colorado and its correlation to the East and the West. U.S. Geol. Survey Prof. Paper 684-A, 19 pp.

Biostratigraphy of Pierre Shale and related sandstone members in Colorado.

Jones, R.L., and W.W. Hay. 1975. Bioliths. In: J.E. Gieseking, Ed., Soil Components. Springer-Verlag, New York, v. 2, p. 481–496.

A compact, east to read, well-illustrated and well-tabled overview by two well-known experts. See Fig. 3.39.

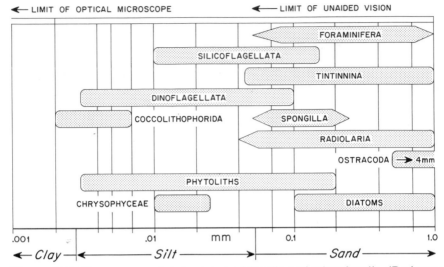

Figure 3.39 Size ranges of major types of fossils and microfossils. (Redrawn from Jones and Hay 1975, Fig. 2)

Müller, German. 1971. Coccoliths: Important rock-forming elements in bituminous shales of Central Europe: Sedimentology, v. 17, p. 119–124.

Chief constituents of carbonate layers in several Tertiary bituminous shales are coccoliths.

Murphy, M.A. 1975. Paleontology and stratigraphy of the Lower Chickabally Mudstone (Barremian–Aptian) in the Ono quadrangle, northern California: Univ. Calif. Publ. Geol. Sci., v. 113, 52pp.

Biostratigraphy using ammonites reveals that there is an angular unconformity at the top of this unit. The Chickabally Member is part of the Budden Canyon Formation, which consists of basinal shales with tongues and lenses of sandstone and conglomerate (cf. Dailey 1973).

Oliver, W.A. Jr., W. de Witt Jr., J.M. Dennison, D.M. Hoskins, and J.W. Huddle. 1969. Correlation of Devonian Rock Units in the Appalachian Basin. U.S. Geol. Survey Oil and Gas Invest. Chart OC-64.

Correlation chart of the Appalachian Basin Devonian rocks, showing lithostratigraphic, time-stratigraphic units, and the biostratigraphic ranges of key brachiopod, conodont, and ammonoid species. Excellent summary of paleontologic markers for a basin-wide shale unit.

Reineck, H-E. 1963. Sedimentgefüge im Bereich der südlichen Nordsee. Abh. Senckenb. Naturf. Ges. 505, 64 pp.

Notable in many ways as an integrated and comprehensive sedimentology study, but also contains a very useful classification of the degree of bioturbation (See Table 3.3 translated from the German).

Table 3.3. Classification of Bioturbation Depending on Primary Bedding Destruction[a]

Grade	Degree of bioturbation (%)	Classification of bioturbation
0	0	No bioturbation
1	1– 5	Sporadic bioturbation traces
2	5–30	Weakly bioturbated
3	30–60	Medium bioturbated
4	60–90	Strongly bioturbated
5	90–99	Very strongly bioturbated, but rest of inorganic bedding still recognizable
6	100	Completely bioturbated

[a] Modified after Reineck (1963, Table 12)

Reiskind, J. 1975. Marine concretionary faunas of the uppermost Bearpaw Shale (Maestrichtian) in eastern Montana and southwestern Saskatchewan. *In:* W.G.E. Caldwell, Ed., The Cretaceous System in the Western Interior of North America. Geol. Assoc. Canada Spec. Paper 13, p. 235–252.

> Isochronous concretions, in a zone about 20 ft thick, occur in silty shales and siltstone in the upper part of the Bearpaw Shale; contain fossils that represent three distinct biofacies; and occur throughout a 28 000 sq. m. region. See Waage (1964) and Colton (1967).

Rhoads, D.C. 1970. Mass properties, stability, and ecology of marine muds related to burrowing activity. *In:* T.P. Crimes and J.C. Harper, Eds., Trace Fossils. Geol. Jour. Spec. Issue 3, p. 391–406.

> Intensive burrowing of subtidal muds produces granular sediment that contains more than 60% water and is easily resuspended, as well as having different mass properties from those of equivalent unburrowed mud.

Thompson, G.G. 1972. Palynologic correlation and environmental analysis within the marine Mancos Shale of southwestern Colorado: Jour. Sed. Petrol., v. 42, p. 287–300.

> Interesting use of microplankton to define environments within a seemingly homogeneous, thick marine shale. Alternation of near-shore, intermediate, and offshore environments correlate well with transgressive–regressive cycles in correlative sandstones.

Waage, K.M. 1964. Origin of repeated fossiliferous concretion layers in the Fox Hills Formation. *In:* D.F. Merriam, Ed., Symposium on Cyclic Sedimentation. Kansas Geol. Survey Bull. 169, p. 541–563.

> Carbonate concretions, some very richly fossiliferous, occur in a Cretaceous silty clay to clayey sand over at least 1500 sq. mi. in Kansas. Different layers have different fossil assemblages. Strong emphasis on paleoecology. Outstanding opportunity to study their orientation to infer paleocurrents as did Colton (1967). See also Reiskind (1975).

Williams, E.G. 1960. Marine and freshwater fossiliferous beds in the Pottsville and Allegheny Groups of western Pennsylvania: Jour. Paleontol., v. 34, p. 908–922.

> A rather unusual paper describes and maps the faunal associations in shales above four different Pennsylvanian coal beds. Table 3.4 gives rock types, faunas, and environments.

Zangerl, R., and E.S. Richardson Jr. 1963. The Paleoecologic History of Two Pennsylvanian Black Shales: Fieldiana Geol. Mem., v. 4, 352 pp.

> One of the most thorough—and amazing—studies of shales ever made, being a *total description* of two thin, black shale seams overlying Pennsylvanian coal beds in the Illinois basin. Large blocks were taken back to the laboratory for complete study that included petrography and geochemistry, in addition to very complete paleoecologic and paleontologic inventory. Fifty-five plates, including 37 photomicrographs of shale thin sections, and a six-page summary of evidence for supporting conclusions. A study that truly must be read to be appreciated! See Figs. 3.40 and 3.41.

SHALES IN ANCIENT BASINS

Regional, integrated studies of shales in ancient basins are very, very hard to find, so that of the citations below, only a very few really merit this distinction. In many respects there are more integrated regional studies of modern muds than there are of their ancient equivalents. Nonetheless, those below do provide useful insights to how ancient shales have been studied; it is hoped that these citations also will give a better appreciation of the occurrence of shales.

Some more notable observations include:

1. Most thick shale sections in geosynclines are marine and appear to have been associated with distal turbidites and, therefore, deposited in relatively deep water.
2. Thin shales can be very, very widespread on cratons.
3. Thick shales commonly have an internal stratigraphy in which large-scale clinoform structure is important.
4. Attention to the physical, biogenic, and diagenetic sedimentary structures of the minor siltstones, sandstones, and limestones that occur within many shales is very rewarding for environmental reconstruction.
5. To find the correct "mud-facies model", in analogy to the now widely used carbonate and sandstone facies models, the integrated study of the shale and its lateral equivalents and associated beds is vital.
6. It is our impression—and only that—that as a rule, most papers reporting on muds and shales in ancient basins make much less use of the Holocene than do most studies of ancient sandstones and carbonates. What do you think?

Table 3.4. Environmental Faunal Classification

Faunal group	Index fossils	Rock type	Associated genera	Inferred environment
4	Estherids	Shale and silty shale, med. dark gray, laminated with platy siderite.	*Estheria, Leaia, Anthraconauta, Limulus, Lepidoderma, Carbonicola* Plant fossils.	Fresh water
3	Phosphatic brachiopods and pectinids	Shale and silty shale, black to dark gray, fissile to platy, nodular siderite concretions	*Lingula, Dunbarella Aviculopecten, Orbiculoidea, Anthraconauta*	Restricted marine or near-shore marine
2	Calcareous brachiopods	Shale, dark gray with nodular siderite concretions.	*Chonetes, Marginifera Composita*	Marine
1	Calcareous brachiopods, cephalopods, pelecypods, and gastropods	Shale, dark gray to gray black, poorly bedded, soft with thin, dark, limestone lenses	*Marginifera, Chonetes, Mesolobus, Composita, Linoproductus, Meekospira, Sphaerodoma, Euphemites, Lophophyllum, Pseudorthoceras*	Marine

*From Williams (1960, Table 2). Published by permission of the author and the Paleontological Society.
Estheria = *Palaeolimnadiopsis.*
Leaia = *Hemicycloidea.*

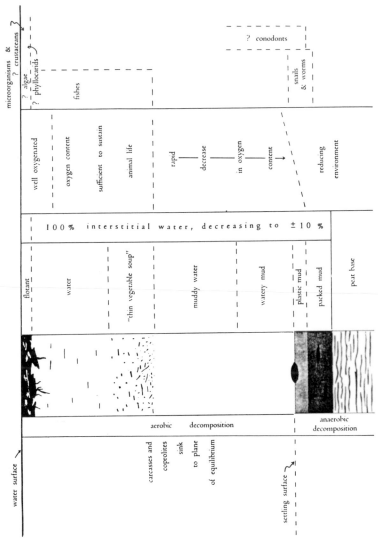

Figure 3.40 Interpretation of habitat and depositional environment of shales in the Mecca and Logan Quarries as residual ponds, based mainly on faunal and lithologic evidence, and some observations on similar modern environments. (Zangerl and Richardson 1963, Fig. 27.)

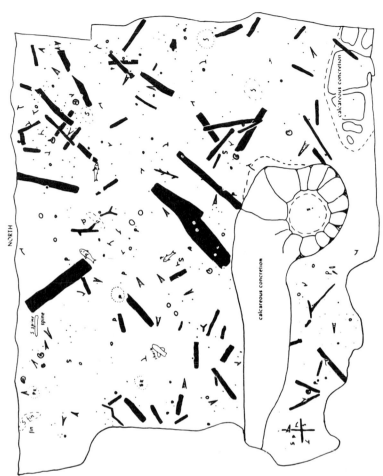

Figure 3.41 Fossils and large concretions in level C, Mecca Quarry shale, Mecca Quarry. Elongated objects have correct orientation. (Zangerl and Richardson 1963, Fig. 34.)

Akhtar, K., and V.K. Srivastava. 1976. Ganurgarh Shale of southeastern Rajasthan, India: A Precambrian regressive sequence of lagoon–tidal flat origin: Jour. Sed. Petrol., v. 46, p. 14–21.

This shale, about 22 m thick, unconformably occurs between an underlying sandstone and an overlying limestone. The shale contains some minor sandstones and siltstones, especially in its lower part. These contain some micro-cross-lamination and ripple lamination and plane beds that indicate a northward paleoflow. These structures, plus mud cracks and raindrop impressions, suggest deposition on a lagoon tidal and supratidal flat.

Andrews, P.B., and A.T. Ovenshine. 1975. Terrigenous silt and clay facies; deposits of the early phase of ocean basin evolution. In: J.P. Kennet, R.E. Houtz, P.B. Andrews, A.R. Edwards, V.A. Gostim, Marta Hajos, M.A. Hampton, D.G. Jenkins, S.V. Margolis, A.T. Ovenshine, and K. Perch-Nielsen, Initial Reports of Deep Sea Drilling Project, v. 29, p. 1049–1063.

Terrigenous clay and silt of two facies, mottled-burrowed and slightly laminated beds, overlie basement in the ocean. The laminites contain shallow-water forams and neritic nannofossils plus some glauconite and are believed to be turbidites whereas the mottled-burrowed facies was deposited in sluggish but oxygenated waters. Both facies are believed to have developed when the continents were sufficiently close together to have inhibited a general oceanic circulation.

Asquith, D.O. 1974. Sedimentary models, cycles, and deltas, Upper Cretaceous, Wyoming, in Rocky Mountain Symposium: Amer. Assoc. Petrol. Geol. Bull., v. 58, p. 2274–2283.

Depositional models, primarily deltaic, are derived from detailed subsurface studies of Upper Cretaceous sandstone and shale facies. Clinoform (falloff) stratification within the Pierre Shale is especially well demonstrated, with diachronous relationships established by correlations of bentonites and other isochronous features in the shale. Very good illustrations. See Fig. 3.42.

Arthur, M.A., and S.O. Schlanger. 1979. Cretaceous "oceanic anoxic events" as causal factors in development of reef-reservoired giant oil fields. Amer. Assoc. Petrol. Geol. Bull. v. 63, p. 870–885.

Special events in the Cretaceous produced organic-rich and thick shales which became source rocks for many of the world's Mesozoic shallow-water carbonate complexes. Authors' Figure 6 (not shown) effectively tells their story.

Berger, W.H., and U. von Rad. 1972. Cretaceous and Cenozoic sediments from the Atlantic Ocean. In: D.E. Hayes and A.C. Pimm, Eds., Initial Reports of the Deep Sea Drilling Project. Natl. Sci. Foundation, Washington, D.C., v. 14 (Leg 14), p. 787–954.

Discussion of all types of sediments, with argillaceous ones prominent. Photographed cores of dark carbonaceous shales and other mudstones on Plates 44–48 (not shown). In the long text there are scattered discussions, by age, of argillaceous sediments with a chief summary on pp. 881–883. Compare dis-

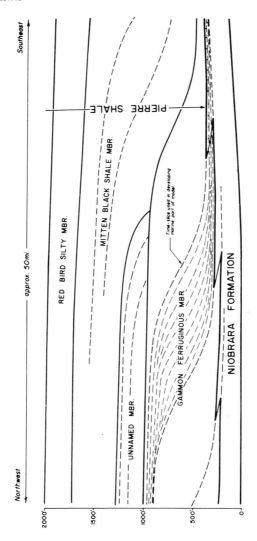

Figure 3.42 Northwest-southeast stratigraphic cross section in eastern Powder River basin. (Asquith 1974, Fig. 2; published by permission of the author and the American Association of Petroleum Geologists.)

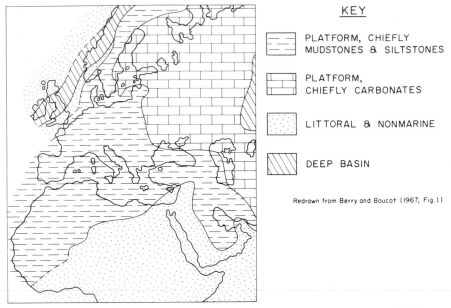

Figure 3.43 Vast extent of Silurian shales in parts of Europe, Asia, and Africa. (Redrawn from Berry and Boucot 1967, Fig. 1.)

cussion of black pyritic shales in Cenomanian (p. 881) with that of Ewing and Hollister (1972).

Berry, W.B.N., and A.J. Boucot. 1967. Pelecypod–graptolite association in the Old World Silurian: Geol. Soc. Amer. Bull., v. 78, p. 1515–1522.

A pelecypod–graptolite association in dark radioactive black shales, including a few dark limestones and some sandstones, occurs over a vast area of northern Africa and southern Europe into the Middle East; thickness ranges from 50 to 300 m. Along its southern edge in Africa, it interdigitates with sandstones. See Fig. 3.43.

Berry, W.B. N., and Pat Wilde. 1978. Progressive ventilation of the oceans —An explanation for the distribution of Lower Paleozoic black shales: Amer. Jour. Sci., v. 278, p. 257–275.

Speculation about the abundance of black shales in Lower Paleozoic and Precambrian chiefly involves photosynthesis, glaciation and their effect on ventilation of the ocean.

Byers, C.W., and D.W. Larson. 1979. Paleoenvironments of Mowry Shale (Lower Cretaceous), Western and Central Wyoming: Amer. Assoc. Petrol. Geol. Bull., v. 63, p. 354–375.

Three distinct facies consisting of laminated mudstone (lethal to life and deeper than 150 m), bioturbated mudstone (many organisms, 15 to 150 m), and bioturbated sandstones (many organisms in agitated water), were deposited on a paleoslope of 0°03'. Facies recognizable on wire line logs.

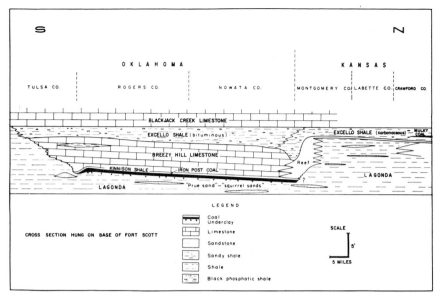

Figure 3.44 Diagrammatic south-north cross section of sedimentary facies of Excello Shale and adjacent strata, southeastern Kansas and northeastern Oklahoma. Vertical scale in feet. (Cassidy 1968, Fig. 4; published by permission of the author and the American Association of Petroleum Geologists.)

Cassidy, M.M. 1968. Excello Shale, northeastern Oklahoma: Clue to locating buried reefs: Amer. Assoc. Petrol. Geol. Bull., v. 52, p. 295–312.

> Interesting use of organic geochemistry to define facies in a seemingly homogeneous shale. This 6-ft shale unit outcrops in four states and is divisible into a northern humic facies and a southern bituminous facies. See Fig. 3.44.

Christensen, L., S. Fregerslev, A. Simonsen, and J. Thiede. 1973. Sedimentology and depositional environment of Lower Danian Fish Clay from Stevns Klint, Denmark: Bull. Geol. Soc. Denmark, v. 22, p. 193–212.

> Detailed chemical and mineralogic study of a thin (less than 20 cm) bed of dark shale containing abundant fish remains. The shale seems to have been deposited in very small basins developed on the underlying Cretaceous chalk.

Cline, L.M. 1966. Late Paleozoic rocks of Ouachita Mountains, a flysch facies. In: Kansas Geol. Soc. Guide Book, 29th Field Conf. on Flysch Facies and Structure of the Ouachita Mountains, November 1966, pp. 91–111.

> A nice, short synthesis of the geologic history of the late Paleozoic shales and sandstones of the Ouachita Mountains in midcontinent United States, showing the change from shallow marine shelf facies to deep-water flysch facies. *Key idea:* Shelf-to-basin transition with thick shales in basin. See Fig. 3.45.

Collinson, J.D. 1969. The sedimentology of the Grindslow Shales and the Kinderscout Grit: A deltaic complex in the Namurian of northern England: Jour. Sed. Petrol., v. 39, p. 194–221.

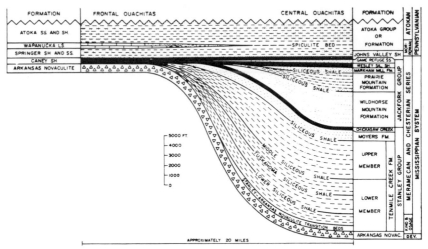

Figure 3.45 Diagrammatic cross section of facies changes and correlations of the Late Mississippian and Early Pennsylvanian formations from the frontal to the central Ouachitas, southeastern Oklahoma; thrust faults eliminated. (Cline 1960.)

The environmental associations proposed in this paper are based almost entirely on field aspects of the sandstones, but the study does provide a good illustration of an important type of sedimentary sequence in which shales are abundant.

Dailly, G. 1975. Some remarks on regression and transgression in deltaic sediments. *In:* C.J. Vorath, E.R. Parker, and D.J. Glass, Eds., Canada's Continental Margins and Offshore Petroleum Exploration. Canadian Soc. Petrol. Geol., Mem. 4, p. 791–820.

Very informative schematic diagrams based on the Niger, Mississippi, and MacKenzie deltas, diagrams that emphasize shale as much as sand and sandstone. Excellent. See Figs. 3.46 and 3.47.

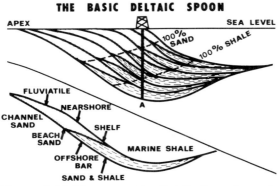

Figure 3.46 Diagrammatic distribution of sand and shale and sedimentary environments in a delta. (Dailly 1975, Fig. 7; published by permission of the author and the Canadian Society of Petroleum Geologists.)

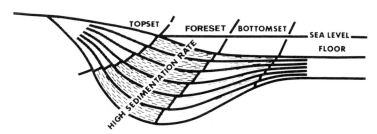

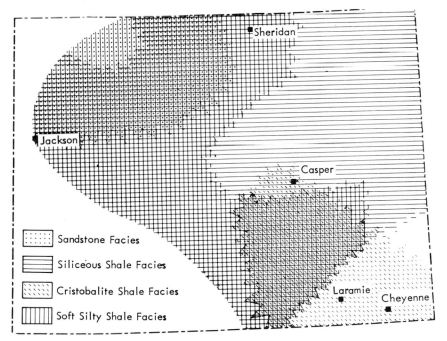

Figure 3.47 Topset, foreset, and bottomset beds of a delta. (Dailly 1975, Fig. 8; published by permission of the author and the Canadian Society of Petroleum Geologists.)

Davis, J.C. 1970. Petrology of Cretaceous Mowry Shale of Wyoming: Amer. Assoc. Petrol. Geol. Bull., v. 54, p. 487–502.

Mineral distribution patterns (clays, feldspar, and quartz) and other evidence suggest that this widespread shale is a transgressive deposit. One of the very few regional studies of shales. Many maps. A possible model study? See Fig. 3.48.

Figure 3.48 Major facies within Cretaceous Mowry Shale of Wyoming. Facies patterns overlap in areas where more than one is present within section. (Davis 1970, Fig. 3; published by permission of the author and the American Association of Petroleum Geologists.)

Dickas, A.B., and J.L. Payne. 1967. Upper Paleocene buried channel in Sacramento Valley, California: Amer. Assoc. Petrol. Geol. Bull., v. 51, p. 873–882.

> More than 95% shale fills a channel over 50 miles long, up to 6 miles wide, and as thick as 2015 ft. Believe down-to-basin faulting created the conditions for channel erosion and that the channel was filled by slumping and turbidity currents. The shale fill is believed to have compacted to about 35%–60% of its original volume. Figure 6 shows a method for computing the relative compaction of channel shale beds. See Hoyt (1959) for another ancient channel fill of shale and Moore and others. (1976) for one possible mechanism of filling.

Drake, A.A. Jr., and J.B. Epstein. 1967. The Martinsburg Formation (Middle and Upper Ordovician) in the Delaware Valley, Pennsylvania–New Jersey. U.S. Geol. Survey Bull. 1244-H, 16 pp.

> Detailed mapping of ten $7\frac{1}{2}$ minute quadrangles shows that the Martinsburg Formation has a total thickness of between 9800 and about 12 800 ft and is divisible into three members: a lower thin-bedded claystone-slate, a middle greywacke turbidite, and an upper thick-bedded claystone-slate. Useful to show how thick shale sections can be associated with deep-water sands.

Dumitriu, M., and C. Dumitriu. 1968. A statistical model of the black shales furrow—Eastern Carpathians: Rev. Roum. Geol. Geophys. Geogr. Ser. Geol., v. 12, p. 99–107.

> Trend surface analysis is applied to sole mark orientation in a ferruginous horizon of black shales (Barremian–Lower Albian) and reveals very uniform orientation over a distance of some 240 km.

Einsele, G., ünd R. Mosebach. 1955. Zur Petrographie, Fossilerhaltung und Entstehung der Gesteine des Posidonienschiefers im Schwäbischen Jura. Neues Jb. Geol. u. Palöntol. Abh. 101, p. 319–430.

> An early post-World War II comprehensive study. Particularly notable are the descriptions of the bedding sequences (Schichtfolgen), the comprehensive paleontology, the discussion of porosity, and geode and concretion morphology, plus the integration of all the different evidence in a well-organized, easy to see diagram (Beilage 2, not shown); 18 conclusions.

Ewing, J.I., and C.H. Hollister. 1972. Regional aspects of deep-sea drilling in the western North Atlantic. In: C.D. Hollister, J.I. Ewing and others, Initial Reports of Deep Sea Drilling Project 11 (Leg 11). Natl. Sci. Foundation, Washington, D.C., p. 951–973.

> Early Cretaceous (Barremian–Cenomanian) dark clay (Horizon A* to Horizon β), zero to more than 500 m thick, is believed to be present over most of North Atlantic; is a very dark (sometimes coal black) color, with much organic matter (p. 967); and appears to have been deposited below the carbonate compensation depth (p. 971). See Berger and von Rad (1972).

Ferm, J.C., and R.A. Melton. 1977. A Guide to the Cored Rocks in the Pocahontas Basin. Carolina Coal Group, Geol. Dept., Univ. South Carolina, Columbia, S.C., 92 pp.

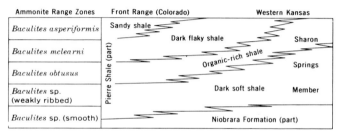

Figure 3.49 Diagram illustrating eastward and upward migration of organic-rich shale unit of Sharon Springs Member of Pierre Shale. (Gill and others 1972, Fig. 9.)

A color picture key to the Pennsylvanian rocks of the Pocahontas Basin in parts of Kentucky, Virginia, West Virginia, and Tennessee, eight major rock types being recognized. Includes a code for rock description and many useful plates, each with photographs of the different rock types. Excellent documentation of shales and their relationships to the sandstones of coal measures.

• Foscolos, A.E., and D.F. Stott. 1975. Degree of Diagenesis, Stratigraphic Correlations and Potential Sediment Sources of Lower Cretaceous Shale of Northeastern British Columbia. Canadian Geol. Survey Bull. 250, 46 pp.

Illustration of use of mineralogy and geochemistry in an attempt to define internal stratigraphy for a thick shale sequence. Fifteen formations named from scattered outcrops were grouped into three separate provenance categories, and a number of correlations proposed.

Gill, J.R., and W.A. Cobban. 1961. Stratigraphy of Lower and Middle Parts of the Pierre Shale, Northern Great Plains. U.S. Geol. Survey Prof. Paper 424-D, pp. 185–191.

Stratigraphic study of a shale sequence that passes from an active source area, producing abundant coarse clastics, in the west, through a quiet black mud shelf zone, to a less active source area to the east. Similar to Devonian in Appalachian Basin?

Gill, J.R., W.A. Cobban, and L.G. Schultz. 1972. Stratigraphy and Composition of the Sharon Springs Member of the Pierre Shale in Western Kansas. U.S. Geol. Survey Prof. Paper 728, 50 pp.

A widespread shale with a persistent threefold internal stratigraphy, the middle one being organic rich. These units can be recognized on wire-line logs. Interesting diagrams about sedimentation rates and facies migration based on ammonite zones, pp. 16–20, plus some schematic maps. Also some clay mineralogy and a goodly amount of trace element study. Compare with Chattanooga. See Fig. 3.49. Compare with Asquith (1974).

Glick, E.E. 1975. Arkansas and northern Louisiana. In: E.D. McKee and E.J. Crosby Coordinators, Paleotectonic Investigations of the Pennsylvanian System in the United States. Part I: Introduction and Regional Analysis of the Pennsylvanian System. U.S. Geol. Survey Prof. Paper 853-I, p. 157–175.

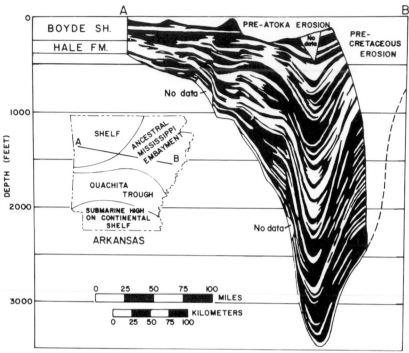

Figure 3.50 Distribution and dominance of Pennsylvanian shale (black) in Ouachita Trough from shelf to basin. (Glick 1975, Figs. 37, 38, and 39.)

Notable for its documentation and geologic history of the Ouachita Trough during the Pennsylvanian. See Fig. 3.50.

Goodson, J.L., Ed. 1964. Shale Environments of the Mid-Cretaceous Section Central Texas—A Field Guide. Baylor Univ. Geol. Sco., Waco, Texas, 77 pp.

The only publication using the phrase "shale environments" that we are aware of has four articles and a road guide, plus 18 illustrations. Aspects studied include petrography, microfauna, clay mineralogy, and subsurface lithofacies, plus sedimentary structures. Certainly a model for future field trips elsewhere.

Gray, H.H. 1972. Lithostratigraphy of the Maquoketa Group (Ordovician) in Indiana. Indiana Geol. Survey Spec. Rept. 7, 31 pp.

This upper Ordovician unit, mostly shale, varies from 1000 to 200 ft thick across Indiana and is a marine clastic wedge derived from the east. Good example of use of wire-line logs for study of a shaly unit. Numerous maps. See Figs. 3.51–3.53

• Greiner, H.R. 1962. Facies and sedimentary environments of Albert Shale, New Brunswick: Amer. Assoc. Petrol. Geol. Bull., v. 46, p. 219–234.

Mississippian lake deposit typically has red beds above and below and was once mined for oil shale. Trends of wave ripples parallel shoreline as inferred

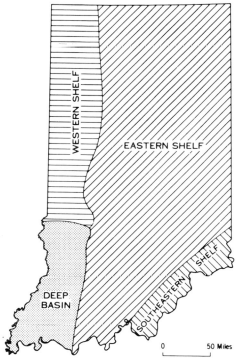

Figure 3.51 Inferred depositional provinces of the Maquoketa Group in Indiana. (Gray 1972, Fig. 4.)

from facies patterns. Hydrocarbon source is believed to have been mostly abundant palaeoniscid fishes.

Gutschick, R.C., M. McLane, and J. Rodriguez. 1976. Summary of Late Devonian–Early Mississippian biostratigraphic framework in western Montana. *In:* Guidebook, The Tobacco Root Geological Society, Field Conf., Montana Bur. Mines and Geol., Spec. Publ. 73, p. 91–124.

Shale, black shale, and carbonate deposition on a craton to miogeosyncline transition, all very well done. Good figures (See Fig. 3.54) and many references.

Hallam, A., and M.J. Bradshaw. 1979. Bituminous shales and oolitic ironstones as indicators of transgressions and regressions. Jour. Geol. Soc., v. 136, p. 157–164.

Document the observation that laminated black shales are usually found at the base of transgressive sequences. By contrast, oolitic ironstones are found in the late stages of regressions. We have observed that Devonian black shale tongues in the Appalachian Basin also seem to mark episodes of rapid transgression followed by a period of slower delta progradation in which bioturbated gray shales and siltstones are deposited.

Hattin, D.E. 1962. Stratigraphy of the Carlile Shale (Upper Cretaceous) in Kansas. Kansas Geol. Survey Bull. 156, 155 pp.

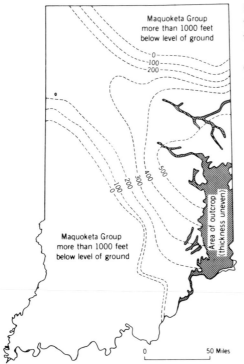

Figure 3.52 Total thickness (in feet) of principal shale beds in the Maquoketa Group less than 1000 feet below the general level of the ground. (Gray 1972, Fig. 15.)

Comprehensive stratigraphy, plus reconstruction of depositional environment and paleogeography. The Carlile and its equivalents (Colorado, Cody, Benton, and Mancos) are believed to represent the regressive half of the first major Late Cretaceous cycle in the western interior United States region. Useful as a model.

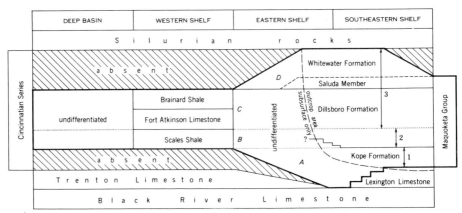

Figure 3.53 Complex nomenclature needed to relate shaly basin to bordering carbonate shelf. (Gray 1972, Fig. 5.)

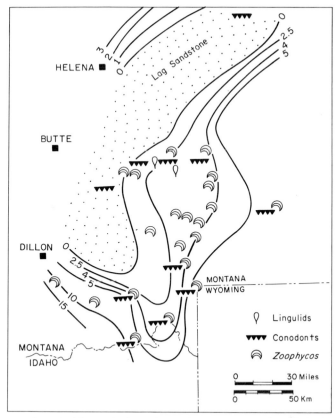

HELENA ■

BUTTE ■

Log Sandstone

DILLON ■

MONTANA
WYOMING

MONTANA
IDAHO

♀ Lingulids

▼▼▼ Conodonts

🐚 *Zoophycos*

0   30 Miles

0   50 Km

Figure 3.54 Pinchout of a thin black shale as it overlaps a craton. Isopachs in meters. (Redrawn from Gutschick and others 1976, Fig. 8-14.)

Hallin, D.E. 1965. Stratigraphy of the Graneros Shale (Upper Cretaceous) in Central Kansas. Kansas Geol. Survey Bull. 178, 83 pp.

A careful, fully integrated, and comprehensive stratigraphic study of a thin (70 m), but very widespread, shale extending from Arizona to southern Canada and beyond. Discussion of its paleontology, tracks and trails, petrology, paleo-ecology, and provenance, plus paleogeography, all nicely summarized in 10 conclusions. More shales need to be studied in this manner. Compare with Davis (1970), Pelzer (1966) and Gill and others (1972). See Fig. 3.55.

Heckel, P.H. 1972. Ancient shallow marine environments. *In:* J.K. Rigby and W.K. Hamblin, Eds., Recognition of Ancient Sedimentary Environments. Soc. Econ. Paleontol. Mineral. Spec. Publ. 16, p. 226–286.

Discussion of environments of deposition of shale and associated rock types on platforms offers possible models, black shale included (pp. 255–264). Many references, but not too many to shale. See Figs. 3.56 and 3.57.

Heckel, P.H. 1977. Origin of phosphatic black shale facies in Pennsylvanian

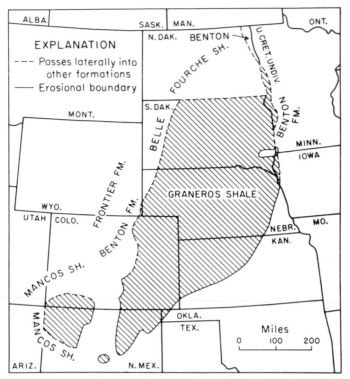

Figure 3.55 Widespread regional distribution of the Graneros Shale and nomenclature of laterally contiguous units. (Hattin 1965, Fig. 4.)

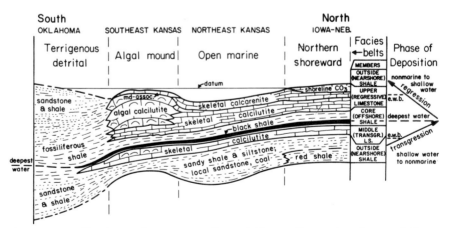

Figure 3.56 North-south cross section of Upper Pennsylvanian cyclothem and interpretation (Heckel 1977, Fig. 4; published by permission of the author and the American Association of Petroleum Geologists.)

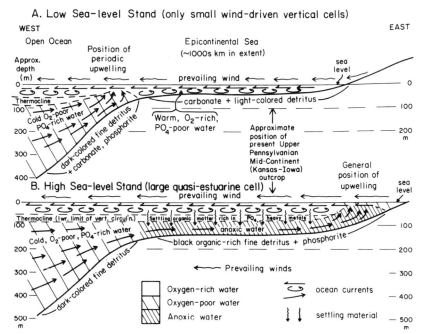

Figure 3.57 Inferred circulation patterns in west-facing tropical epicontinental sea. (Heckel 1977, Fig. 5; published by permission of the author and the American Association of Petroleum Geologists.)

cyclothems of Mid-Continent North America: Amer. Assoc. Petrol. Geol. Bull., v. 61, p. 1045–1068.

The black shales of Pennsylvanian cyclothems are used to develop a general model of their deposition, at least on cratons, that may be widely applicable, as suggested by Figs. 3.56–3.58. Excellent.

Hesse, R. 1975. Turbiditic and non-turbiditic mudstone of Cretaceous flysch sections of the East Alps and other basins: Sedimentology, v. 22, p. 387–416.

Comprehensive criteria in Table 3 to differentiate between turbidite and nonturbidite mudstones in Mesozoic and Cenozoic turbidite basins. Significant because many thick basinal shales are associated with distal turbidites. To what extent can these and related criteria be applied to differentiate subenvironments of shales in nonturbidite basins?

Hoover, K.V. 1960. Devonian–Mississippian Shale Sequence in Ohio. Ohio Geol. Survey Inf. Circ. 27, 154 pp.

Good bibliographic review of the geology of an important shale unit as of 1960. Faunal lists, isopach and structure maps, paleogeographic maps, and annotated bibliography.

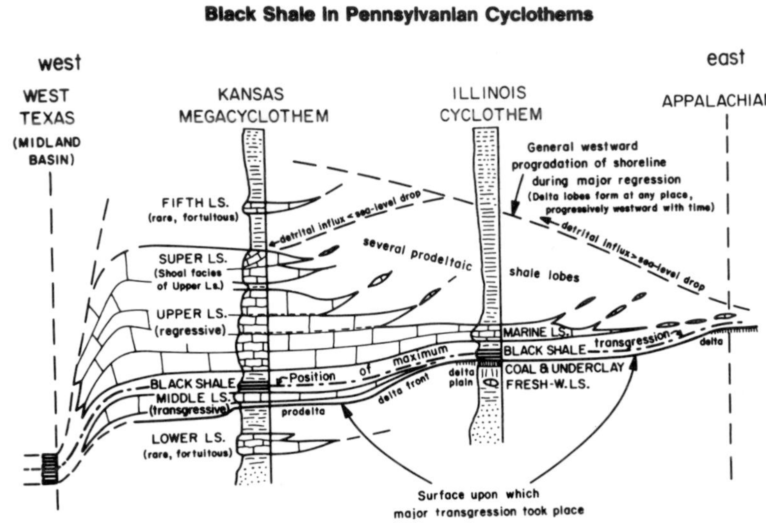

Black Shale In Pennsylvanian Cyclothems

Figure 3.58 Idealized model of Pennsylvanian sedimentation in response to eustatic transgressions and regressions on a wide craton and a bordering trough. (Heckel 1977, Fig. 7; published by permission of the author and the American Association of Petroleum Geologists.)

Hoyt, W.V. 1959. Erosional channel in the Middle Wilcox near Yoakum, Lavaca County, Texas: Trans. Gulf Coast Assoc. Geol. Socs. v. 9, p. 41–50.

This Middle Wilcox (Tertiary) channel is over 60 miles long, averages 5 miles in width, has maximum depth of about 3000 ft, and is filled almost entirely by shale. See also Dickas and Payne (1967).

Kauffman, E.G. 1969. Cretaceous marine cycles of the western interior: Mountain Geol., v. 6, p. 227–245.

Classic, diagramatic representation of lithologies associated with transgression and regression on a stable craton (Fig. 4, not shown) shows lateral equivalents of a mud.

Lewis, T.L., and J.F. Schwietering. 1971. Distribution of the Cleveland black shale in Ohio: Geol. Soc. Amer. Bull., v. 82, p. 3477–3483.

Study of a single unit within a thick shale sequence. Isopach maps show marked pinching and swelling along strike that may be related to turbidite fans from the east from the Catskill Delta. See Figs. 3.59 and 3.60.

Lineback, J.A. 1970. Stratigraphy of the New Albany Shale in Indiana: Indiana Geol. Surv. Bull., v. 44, 73 pp.

Modern stratigraphic analysis, subsurface and outcrop. Suggests a floating algal mat as the source of the organic matter. Very well done.

Ludwig, G. 1964. Divisão Estratigráfico-Faciológica do Paleozóico da Bacia Amazônica. Petrobrás, Secão de Explorcão de Petróleo, Publ. 1, 55 pp.

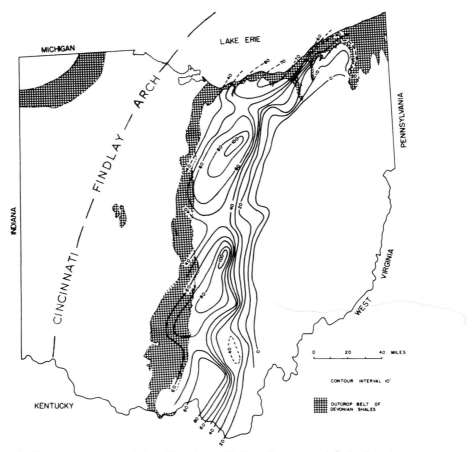

Figure 3.59 Isopach map of the Cleveland Shale. (Lewis and Schwietering 1971, Fig. 3; published by permission of the authors and the Geological Society of America.)

Early effort based primarily on cores and GR/N logs to interpret environmentally the sedimentary structures and lithology of argillaceous sediments in the subsurface of a Middle Devonian shale over 1000 m thick. Eleven plates of sedimentary structures and bedding, plus an interesting way to relate lithology and sedimentary structures to wire-line logs graphically.

MacKenzie, W.S. 1972. Fibrous calcite, a Middle Devonian geologic marker, with stratigraphic significance, District of MacKenzie, Northwest Territories: Canadian Jour. Earth Sci., v. 9, p. 1431–1440.

Thin, fibrous calcite beds, with maximum thickness of 7 cm and having cone-in-cone structure, occur in a widespread black shale over a 24 000 km² area in the District of MacKenzie and are, therefore, excellent marker beds. Believes markers developed close to sediment–water interface. The associated shales are very similar to the Devonian black shales of the Appalachians.

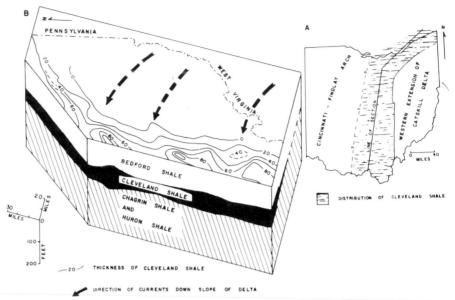

Figure 3.60 Facies relationship of the Cleveland Shale with enclosing rock masses. (Lewis and Schwietering 1971, Fig. 4A and B; published by permission of the authors and the Geological Scoiety of America.)

McGugan, A. 1965. Occurrence and persistence of thin shelf deposits of uniform lithology: Geol. Soc. Amer. Bull., v. 76, p. 125–130.

> Defines a *persistence factor* that should be used more commonly to help define the lateral continuity of shales, which is commonly very great (cf. Berry and Boucot 1967).

Meyers, R.L., and D.C. van Siclen. 1964. Dynamic phenomena of sediment compaction in Matagorda County, Texas: Trans. Gulf Coast Assoc. Geol. Socs., v. 14, p. 241–252.

> Relates cyclical variation in densities of deeply buried shales in the Texas Gulf Coast to interstitial fluid pressures, salinity, and permeability. Discusses the origin of low-density–high-pressure shales. See Fig. 3.61.

Morris, R.C. 1974. Carboniferous rocks of the Ouachita Mountains, Arkansas: A study of facies patterns along the unstable slope and axis of a flysch trough. *In:* G. Briggs, Ed., Carboniferous Rocks of the Southeastern United States. Geol. Soc. Amer. Spec. Paper 148, p. 241–280.

> See Fig. 3.62 for shale in a geosyncline.

Musgrave, A.W., and W.G. Hicks. 1968. Outlining shale masses by geophysical methods. *In:* J. Braunstein, and G.D. O'Brien, Eds., Diapirism and Diapirs. Amer. Assoc. Petrol. Geol. Mem. 8, p. 122–136.

> Shale masses have three properties: (1) low-velocity sound transmission at low

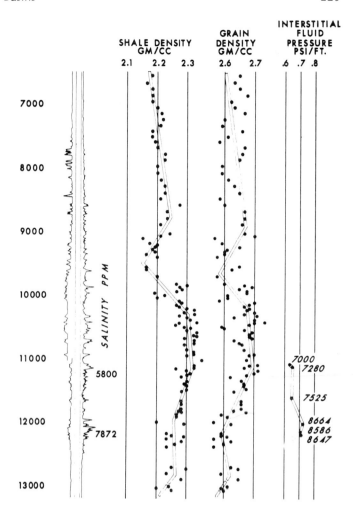

Figure 3.61 Vertical variation of shale and grain density and interstitial fluid pressure for Tenneco No. 1 M. L. Davis Gas UT.1 well. (Meyers and van Siclen 1964, Fig. 3; published by permission of the authors and the Association of Gulf Coast Geological Societies.)

density; (2) low resistivity; and (3) high fluid pressure, all of which appear to result from shale's high porosity and low permeability (at least in Mesozoic-Tertiary shales of Louisiana). See Fig. 3.63.

Nixon, R.P. 1973. Oil source beds in the Cretaceous Mowry Shale of northwestern interior United States: Amer. Assoc. Petrol. Geol. Bull., v. 57, p. 136–161.

Excellent regional study of a shale. Emphasizes the use of organic measurements to define areas of best oil source bed development. It was found that burial to at least 7000 ft is required to produce appreciable quantities of petroleum hydrocarbons in this basin.

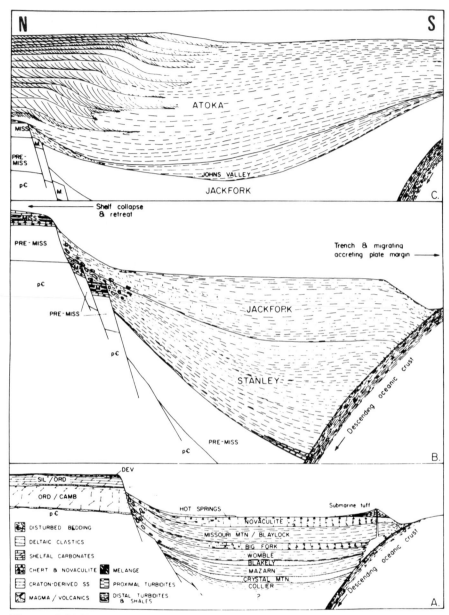

Figure 3.62 Restored diagrammatic cross sections of the Ouachita Trough: C) after Atoka deposition, but before Ouachita orogeny, B) end of deposition of Jackford and A) earliest Stanley deposition. Notable features include shelf to basin transition, growth faults along north side of Ouachita Trough and greater proportion of shale in trough than on shelf. (Morris 1974, Fig. 27; published by permission of the author and the Geological Society of America.)

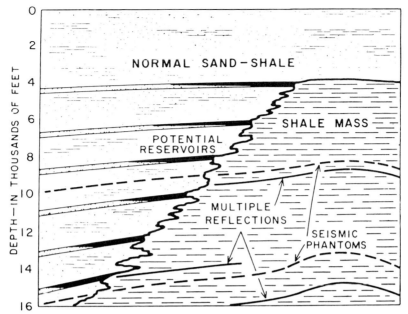

Figure 3.63 Hydrocarbons trapped by shale mass. Figure shows necessity of distinguishing between multiple reflections, which may occur as predominant lineups within shale mass, and true reflections in normal sand shale sequence. (Musgrave and Hicks 1968, Fig. 25; published by permission of the authors and the American Association of Petroleum Geologists.)

• Picard, M.D., and L.R. High Jr. 1968. Sedimentary cycles in the Green River Formation (Eocene), Uinta Basin, Utah: Jour. Sed. Petrol., v. 38, p. 378–383.

A short summary of the cyclic facies and their depositional environments in the Parachute Creek Member of the Green River Shales. The kerogen shales are most common in the deep-water lacustrine cycles.

Pryor, W.A., and H.D. Glass. 1961. Cretaceous–Tertiary clay mineralogy of the Upper Mississippi Embayment: Jour. Sed. Petrol., v. 31, p. 38–51.

One of the first regional studies relating clay mineralogy of stratigraphic units to depositional environments; integration of sedimentary structures, heavy minerals, thin sections, macrofauna, micropaleontology, and X-ray diffraction analyses. Shows that kaolinite is associated with a fluvial facies, and montmorillonite with a marine facies. See Figs. 3.64 and 3.65.

Schermerhorn, L.J.G. 1974. Late Precambrian mixtites: Glacial and/or nonglacial? Amer. Jour. Sci., v. 274, p. 673–824.

A long essay on "tillites" consisting of more or less sandy mudstones containing pebbles and boulders, some of which cover very wide areas in the Precambrian and in the Phanerozoic. Annotated here largely to point out the vast range of sediment types included under the heading of "argillaceous sediments." Almost

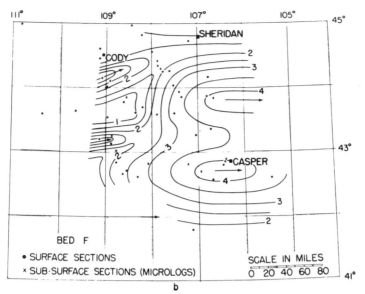

b

Figure 3.67 Total isopach maps of bed F. (Slaughter and Earley 1965, Fig. 6; published by permission of the authors and the Geological Society of America.)

synthesis of surface and subsurface stratigraphy, thin sections, grain mounts, size analyses, and X-ray diffraction analyses to determine correlations, mineralogic origin, source area, and geologic history of thin, widespread bentonitic marker beds. See Figs. 3.67 and 3.68.

Spencer, A. M., Ed. 1974. Mesozoic–Cenozoic Orogenic Belts. Scottish Acad. Press for the Geol. Soc., Edinburgh and London, Spec. Publ. 4, 809 pp.

Figure 3.68 East-west trends of bentonites and Cretaceous batholiths: the batholiths in black and bentonites by open ovals. (Slaughter and Earley 1965, Fig. 23; published by permission of the authors and the Geological Society of America.)

Looking for a source book to guide you to thick Mesozoic and Cenozoic geosynclinal shales? This massive, systematically edited volume is it, with Figs. 12, 13, 14, and 15 (not shown) providing a summary of shale occurrence, as well as the many detailed tables of the 45 orogenic belts described in the text. Remarkable collection of data.

Stehli, F.G., W.B. Creath, C.F. Upshaw, and J.M. Forgotson Jr. 1972. Depositional history of Gulfian Cretaceous of East Texas Embayment: Amer. Assoc. Petrol. Geol. Bull., v. 56, p. 38–67.

A basin analysis, from subsurface information, of a predominantly shaly sequence. Uses a few simple lithologic measurements to construct a variety of basin-wide maps, which are then used to make the interpretations. One of the few articles mapping and interpreting shale color and thickness. Black shale is believed to have been deposited in an open ocean seaward of Angelina–Caldwell flexure in well-oxygenated water, according to microfauna. Shale is black where material is finest and least permeable because it is farthest from source. Also considers and rejects possibility that shale color is diagenetic (cf. McCrossan 1955). Relevant model for the Upper Devonian black shales of Appalachian Basin? See Fig. 3.69.

Stel, J.H. 1975. The influence of hurricanes upon the quiet depositional conditions in the Lower Emsian La Vid Shales of Colle (N.W. Spain): Leidse Geol. Med., v. 49, p. 475–486.

Generally without fossils and very fissile, this shale is believed to have been deposited in an abiotic environment well below wave base, except when storms altered wave base and allowed a pioneer fauna to develop. Lithologic and bedding contrasts support this interpretation. How many other shales record similar episodes?

Sundelius, H.W. 1970. The Carolina slate belt. In: G.W. Fisher, F.J. Pettijohn, J.C. Reed Jr., and K.N. Weaver, Eds., Studies of Appalachian Geology. Wiley-Interscience, New York, p. 351–367.

Summary of the deformed, chiefly argillaceous, fill (about 9 km thick and now mostly in the greenschist facies of metamorphism) of a Paleozoic eugeosyncline. The argillaceous fill has been termed slate, volcanic slate, shale, mudstone, argillite, and siltstone and is interbedded with volcanic wacke and siltstone, quartzites, limestone, and volcanics. Abundant petrographic and chemical data, plus some discussion of the significance and potential for stratigraphic correlations within the belt. This paper is one possible model of how to study a deformed, low-rank metamorphic, shale-filled basin.

Sutton, R.G., Z.P. Bowen, and A.L. McAlester. 1970. Marine shelf environments of the Upper Devonian Sonyea Group of New York: Geol. Soc. Amer. Bull., v. 81, p. 2975–2992.

A shelf-to-basin transition in a delta complex combining stratigraphy with the study of fossil communities—with black shales at the distal end. See Figs. 3.70 and 3.71.

Thompson, A.M. 1972. Shallow Water Distal Turbidites in Ordovician Flysch, Cen-

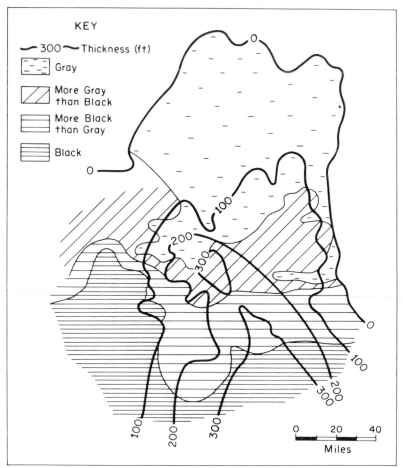

Figure 3.69 Thickness and color of a shale. More maps such as this are needed! (Redrawn from Stehli and others 1972, Fig. 21.)

tral Appalachian Mountains, U.S.A. Internatl. Geol. Cong., 24th Sess., Sec. 6, ed. J. Gill, Montreal, Canada, p. 89–99.

Fig. 3.72 shows diagramatically the relation of flysch shale to flysch sandstone in a prograding flysch basin. See also Morris (1974) for a transverse cross section.

Tweedie, K.A.M. 1968. The stratigraphy and sedimentary structures of the Kimberley Shales in the Evander Goldfield, Eastern Transvaal, South Africa: Geol. Soc. Trans. South Africa, v. 71, p. 235–256.

A remarkable subsurface study of a thin, 110 ft, shale with minor siltstones in a small area, about 4 miles, in a gold mine in the Precambrian Witwatersrand. Very detailed descriptions of slump structures, a paleocurrent map, ripples, and groove casts; descriptions of concretions, mud cracks, and metamorphism; and

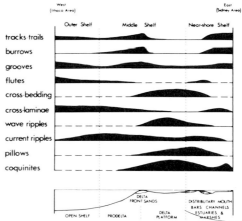

Figure 3.70 Generalized east west variation of sedimentary structures in rocks of study region. Height of black bars proportional to abundance of structure. Inset shows schematic cross-section of inferred sedimentary environments. (Sutton and others 1970, Fig. 5; published by permission of the authors and the Geological Society of America.)

a section on basin analysis, plus eight chemical analyses. Very few, if any, papers comparable to this one.

Valdiya, K.S. 1970. Simla Slates; The Precambrian flysch of the Lesser Himalaya, its turbidites, sedimentary structures and paleocurrents: Geol. Soc. Amer. Bull., v. 81, p. 451–467.

A shaly flysch about 800 m thick, in which the argillaceous sediments are slates and phyllites with interbedded siltstones and lithic and quartz wackes both of which provide most of the paleocurrent data, two trends being recognized. Of interest chiefly to see how the interbedded sandstones of a shale–slate sequence

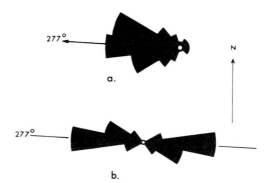

Figure 3.71 Directional structures: 136 measurements of unidirectional indicators and measurements of grooves at 58 localities. (Sutton and others 1970, Fig. 6; published by permission of the authors and the Geological Society of America.)

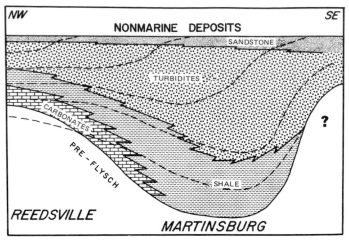

Figure 3.72 Inferred sequence of basin filling in Reedsville-Martinsburg flysch basin. Dashed lines represent successive time lines. Configuration of southeastern margin of Martinsburg basin is hypothetical, but necessitated by a southeastern sediment source. Not to scale. (Thompson 1972, Fig. 8.)

contribute to an improved perception of the origin of the associated shales and slates.

• van Houten, F.B. 1964. Cyclic lacustrine sedimentation, Upper Triassic Lockatong Formation, Central New Jersey and adjacent Pennsylvania. *In:* D.F. Merriam, Ed., Symposium on Cyclic Sedimentation. Kansas Geol. Survey Bull. 169, Part. 2, p. 497–531.

In this Triassic basin there are shales and argillites in excess of 9500 ft thick which have detrital cycles 14–20 ft thick and chemical cycles 8–13 ft thick. Rock types include black shale, platy dark gray shale, carbonate-rich mudstone, and silty mudstone. Good descriptions of lithologies and sedimentary structures. Deposits considered lacustrine.

Villwock, J.A. 1972. Aspectos tectônicos da deposição de folhelhos pretos; Comparação entre a Formação Irati e o "Chattanooga Shale": Pesquisas, n. 1, p. 25–33.

Compares a lower Permian widespread black shale in the Paraná Basin to the Chattanooga—a comparative basin analysis of which we need many more for shales. English abstract and 26 references.

von Gaertner, H.R., H. Kroepelin, H-H. Schmitz, H. Fesser, K. Mädler, H. Jacob, und K. Hoffman. 1968. Zur Kenntnis des nordwestdeutschen Posidonienschiefers. Beihefte Geologischen Jahrbuch, v. 58, 579 pp.

Seven articles on almost all aspects of a widespread Jurassic shale in West Germany—lithology, organic chemistry, paleontology, etc. Seventy-two figures and 100 tables and excellent plates. Classic comprehensive study. English abstract.

Walls, R.A. 1973. Late Devonian–Early Mississippian subaqueous deltaic facies in a

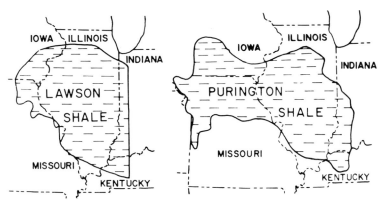

Figure 3.73 Two widespread, thin shales of Pennsylvanian age on the North American craton invite careful integrated sedimentologic and paleoecologic study. (Redrawn from Wanless 1964, Fig. 6 and 8.)

portion of the southeastern Appalachian Basin: Trans. Gulf Coast Assoc. Geol. Socs., v. 23, p. 41–45.

> Subsurface facies study of a gas-bearing black shale sequence ranging in thickness from 1200 to over 4000 ft, plus some petrography and size data on the interbedded siltstones and sandstones. Figure 2 (not shown) is a very helpful correlation table.

Wanless, H.R. 1952. Studies of field relations of coal beds. *In:* Second Conference on Origin and Constitution of Coal. Nova Scotia Dept. Mines and Nova Scotia Res. Found., p. 148–180.

> Chiefly aimed at the geometry of coal beds, but much on clay partings in coals, some of which are very widespread, thin (1 or 2 cm thick), distinctive stratigraphic markers in coal beds. One of the most famous is the Blue Band in the Illinois No. 6 Coal, which may extend for as much as 550 miles or more and the origin of which has never been well explained.

Wanless, H.R. 1964. Local and regional factors in Pennsylvanian cyclic sedimentation. *In:* D.F. Merriam, Ed., Symposium on Cyclic Sedimentation. Kansas Geol. Survey Bull. 169, Pt. 2, p. 593–606.

> Maps and cross sections emphasizing the facies relations and wide lateral extent of the thin shales in Pennsylvanian, cratonic cyclothems, many of which [cf. the Lawson shale of Fig. 3.73] cover more than 50 000 sq. mi. See also Fig. VII-6 of Wilson's (1975) Carbonate Facies in Geologic History (Springer-Verlag, New York, 471 pp.), and Wanless' Fig. 3.74.

Wengerd, S.A., and J.W. Strickland. 1954. Pennsylvanian stratigraphy of Paradox Salt Basin, Four Corners Region, Colorado and Utah: Amer. Assoc. Petrol. Geol. Bull., v. 38, p. 2157–2199.

> Analysis of a basin containing an interesting black shale-evaporite association. Uses isopach and lithofacies maps to determine effects of tectonics on sedimentation. See Woolnough (1937) for a discussion of this association.

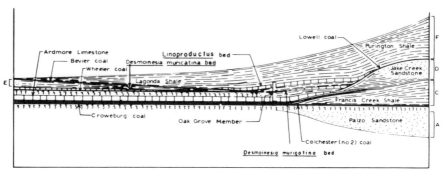

Figure 3.74 Idealized cross section (not to scale) illustrates clastic wedges within the Liverpool cyclothem and its correlatives: (A) Palzo Sandstone, southern Illinois, (B) Browning Sandstone, northern Illinois (not shown), (C) Francis Creek Shale, (D) Jake Creek Sandstone and shale, (E) shale between Wheeler and Bevier coals, and (F) Purington Shale (lower part of wedge). (Wanless 1964, Fig. 7.)

BURIAL HISTORY

It is perhaps in understanding the postdepositional changes, which shales undergo, that the best progress has been made in shale research over the past 10 years. We have divided our references into "Compaction and Structure," the largely physical effects of water loss; "Diagenesis," mostly chemical and mineralogic transformations; and "Thermal History," the application of some of these chemical and mineralogic data to reconstructing the temperature history of a basin. We have also included a few papers on how the amount and composition of shale have varied throughout geologic time.

Compaction and Structure

Compaction in thick shale sequences generates important tectonic movements and expels large amounts of water. Therefore, there is a great deal of current research from both theoretical and practical viewpoints. It is also very important in surficial processes, as discussed under "Environmental and Engineering Geology." Clays do not release water in a simple fashion as overburden pressure increases. Instead, there is a tendency for water expulsion to lag behind burial, leading to a "compaction disequilibrium" that can, on further burial, produce a variety of structural movements. In addition, the fact that undercompacted shales retain their water means that more fluid is available to transport hydrocarbons in the depth range where they are generated.

Two phenomena related to undercompaction have not be adequately integrated into this picture. The first is the role of water released by smectite dehydration, as described in some of the references under "Diagenesis." The second is the development of large salinity variations in some deeply buried pore waters. These are affected by two processes: an anion exclusion, in which Cl⁻ is expelled from the negatively charged shales into the sands, and a membrane exclusion or salt filtering effect, in which dissolved substances are retarded as water passes through the clay.

Barker, C. 1972. Aquathermal pressuring: Role of temperature in development of abnormal-pressure zones: Amer. Assoc. Petrol. Geol. Bull., v. 56, p. 2068–2071.

Shows that a part of the excess pressure found in some thick shales at depth may be caused by the increasing volume of water with increasing temperature (the solids do not expand appreciably). If this volume cannot be released through fluid migration, overpressuring will result. A temperature gradient of 15°C/km or higher is necessary.

Bishop, R.S. 1979. Calculated compaction states of thick abnormally pressured shales. Amer. Assoc. Petrol. Geol. Bull. v. 63, p. 918-933.

Theory plus graphic methods of solution.

Braunstein, J., and G.D. O'Brien, Eds. 1968. Diapirism and Diapirs—A Symposium. Amer. Assoc. Petrol. Geol. Mem. 8, 444 pp.

Twenty papers and an indexed bibliography with four on shale diapirs and mud lumps: recognition on electric logs, Mississippi mud lumps, Tertiary mud diapirs in Texas, and recognition by geophysical methods. Bibliography includes headings on clastic dikes, mud flows and mud volcanoes, mud lumps, and shale intrusions (15 entries). The best *single source* to consult.

Brooner, F.I. Jr. 1967. Shale diapirs of the Lower Texas Gulf Coast as typified by the North LaWard Diapir: Trans. Gulf Coast Assoc. Geol. Socs., v. 17, p. 126–134.

This diapir, described as typical for this region, underlies about 1 sq. mi., has intruded the overlying formation more than 1000 ft., and causes some subsequent formations to thin over it. Distinguishes between piercement and nonpiercement shale diapirs and suggests that the latter are better traps than the former (assuming that migration occurred before piercement); i.e., the timing and extent of intrusion in relation to migration. A large volume of deep-water shale (rapidly deposited?) is believed needed to produce diapirs.

Bruce, C.H. 1973. Pressured shale and related sediment deformation: Mechanism for development of regional contemporaneous faults: Amer. Assoc. Petrol. Geol. Bull., v. 57, p. 878–886.

Thick, rapidly deposited shales sequences are often "undercompacted," that is, their bulk density is too low for normal shales with similar overburden. This instability can lead to faulting, which is therefore caused by sedimentation. Structures formed in this way may be important petroleum reservoirs because large volumes of water move up into them from heated, organic matter-bearing shales. See Fig. 3.75.

Burst, J.F. 1976. Argillaceous sediment dewatering. In: F.A. Donath, Ed., Annual Review of Earth and Planetary Sciences, Annual Reviews, Inc., Palo Alto, Calif., v. 4, p. 293–318.

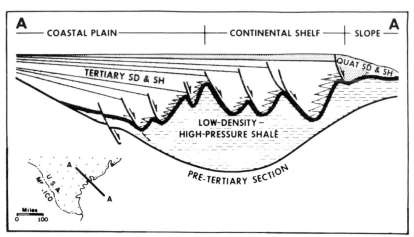

Figure 3.75 Shelf-to-basin transition and low-density, high-pressure shales, which produce deformation in overlying sands. (Redrawn from Bruce 1973, Fig. 1.)

Sediment dewatering, once thought to be solely the result of gravity, is a complicated interaction of kinetic, thermodynamic, and electrochemical processes, especially at greater depths, where clay mineral diagenesis becomes important. Over 70 references.

Chapman, R.E. 1974. Clay diapirism and overthrust faulting: Geol. Soc. Amer. Bull., v. 85, p. 1597-1602.

Marine clays, formed during regression and buried to two or three km, can, if uncompacted and mechanically unstable, form narrow anticlines separated by broad, gentle synclines. The strike of such anticlines should parallel basin margins and, if a slope develops before excess pressure in the shales is dissipated, basinward overthrusting may develop. This interesting theory, still to be fully developed and verified, links the sedimentary filling of a basin to initial deformation. Could it be, in some complexly deformed basins, that the shales started it all?

• Conybeare, C.E.B. 1967. Influence of compaction on stratigraphic analysis: Canadian Petrol. Geol. Bull., v. 15, p. 331–345.

Discussion of how variable compaction, mostly by differential deposition of mud, is a major factor in correlation and interpretation of shaly sections. Worked examples.

Dailly, G.C. 1976. A possible mechanism relating progradation, growth faulting, clay diapirism and overthrusting in a regressive sequence of sediments: Canadian Petrol. Geol. Bull., v. 24, p. 92–116.

A well-illustrated paper that elaborates on the theme of progradation over undercompacted shale, growth faults, clay diapirism, and overthrusting.

Dickey, P.A. 1972. Migration of interstitial water in sediments and the concentration of petroleum and useful minerals. Internatl. Geol. Cong., 24th Sess., Sec. 5, ed. J. Gill, Montreal, Canada p. 3–16.

> A good overview with 12 figures, mostly graphs and diagramatic cross sections, about abnormally high pressures, especially in shales, and the chemical processes associated with compaction. *Key idea:* Overpressure may burst enclosing sediment and resulting fissures may be mineralized or become conduits for petroleum migration. Forty references. See also Powers (1967).

• Fertl, W.H. 1977. Shale density studies and their application. In: G.D. Hobson, Ed. Developments in petroleum geology-1, G.D. Hobson, ed. London, Applied Science Publications Ltd., p. 293-327.

> Probably the best single article, one that is well written and also very well illustrated. Contains 69 references. Start here.

Hamilton, W. 1974. Map of Sedimentary Basins of the Indonesian Region. U.S. Geol. Survey Map I-875B.

> In the short accompanying text one can find good examples of the mobility of thick shales in an actively deforming geosyncline: shale diapirs, and long, narrow, shale cored anticlines growing concurrently with sedimentation. Here deformation has squeezed mostly Tertiary shales into anticlines, some of which have sufficient topographic expression to be capped by modern reefs and shoals.

Hanshaw, B.B., and T.B. Coplen. 1973. Ultrafiltration by a compacted clay membrane; II—Sodium ion exclusion at various ionic strengths: Geochim. Cosmochim. Acta, v. 37, p. 2311-2327.

> A quantitative presentation of Hanshaw's ideas on the production of highly saline subsurface brines through clay membrane filtration. Although this explanation for these brines is the most widely accepted one today, there are quantitative and qualitative problems (see, for instance, Kharaka and Berry 1973). The composition of shale waters may have a strong influence on overpressuring (Schmidt 1973).

Hedberg, H.D. 1974. Relation of methane generation to undercompacted shales, shale diapirs, and mud volcanoes: Amer. Assoc. Petrol. Geol. Bull., v. 58, p. 661–673.

> Generation of methane may contribute to buildup of fluid pressure in shales. If this overpressure is released by diapiric movement of the shale, the methane or other hydrocarbons may be transmitted to reservoir rocks originally remote from the shale. Good review of overpressuring and methane generation, with 99 references.

Higgins, G.E., and J.B. Saunders. 1967. Report on 1964 Chetham Mud Island, Erin Bay, Trinidad, West Indies: Amer. Assoc. Petrol. Geol. Bull., v. 51, p. 55–64.

> Description of a mud diapir which surfaced offshore. At its maximum, this diapir rose 25 ft. above sea level and covered about 10.5 acres. Although it contained some gas, subsurface tectonic deformation is believed to have been responsible.

Katz, H.R. 1975. Ariel Bank off Gisborne; an offshore Late Cenozoic structure, and the problem of acoustic basement on the east coast, North Island, New Zealand: New Zealand Jour. Geol. Geophys., v. 18, p. 93–107.

> Example of geophysical problems caused by thick overpressured shales. In this case, Lower Tertiary (?) mudstones have apparently produced diapirs and mud-cored anticlines in younger sediments, but their internal structure cannot be seen seismically, nor can the zone from which they have risen, because of high fluid pressures.

Kharaka, Y.F., and F.A.F. Berry. 1973. Simultaneous flow of water and solutes through geological membranes; I—Experimental investigation: Geochim. Cosmochim. Acta, v. 37, p. 2577–2603.

> An attempt to verify Hanshaw's model (see Hanshaw and Coplen 1973) for the development of subsurface brines by clay membrane filtration. The qualitative behavior of different ions is reported, an important extension of earlier work, but cannot be predicted from the model because several parameters in Hanshaw's equations have yet to be determined. The change in behavior of the ions with changing flow rate suggests that equations from chromatography theory may prove useful.

Magara, K. 1975. Reevaluation of montmorillonite dehydration as cause of abnormal pressure and hydrocarbon migration: Amer. Assoc. Petrol. Geol. Bull., v. 59, p. 292–302.

> Calculations are presented to show that montmorillonite dehydration by itself cannot explain overpressuring in shales, but may be a contributing factor in some basins. Introduces the useful term "compaction Disequilibrium" to describe the state of overpressureing or undercompaction in shales.

Magara, K. 1976. Water expulsion from clastic sediments during compaction: Direction and volumes: Amer. assoc. Petrol. Geol. Bull., v. 60, p. 543–553.

> Proposes two common models for expulsion of water from shales: dominantly horizontal flow from center to edges in interbedded sandstone–shale sequences, and concomitant vertical flow in thick shale sections unbroken by sandstones. Combinations of these two types are also recognized. Interesting graphs, especially of volume of water *expelled* per volume of shale versus depth.

McCrossan, R.G. 1955. Colour variations in Ireton Shale of Alberta: Canadian Petrol. Geol. Bull., v. 5, p. 48–51.

> Believes that the color of these shales is more closely related to their depth of burial and porosity than to original sedimentation—certainly an idea that deserves much more attention.

Meade, R.H. 1966. Factors influencing the early stages of compaction of clays and sands—A review: Jour. Sed. Petrol., v. 36, p. 1085–1101.

> A well-illustrated essay review primarily aimed at compaction in clays. Factors chiefly affecting early compaction include rate of deposition, grain size, and clay mineralogy (smectite retains the most water). Minor additional factors

include organic matter and temperature. There is also some discussion of clay fabric and compaction of sands.

Mifflin, M.D. 1970. Mudlumps and suggested genesis in Pyramid Lake, Nevada. *In:* Hydrology of Deltas. Proc. Internatl. Assoc. Scientific Hydrology, 1969 Bucharest Symposium, v. 1, p. 75–88.

Mudlumps similar to those of the Mississippi Delta are forming in Pyramid Lake. This well-documented case suggests that for mud lumps to form, there must be rapid progradation of sediment over offshore muds, which are then "overpressured." Intrusion occurs when the pore pressure of the mud exceeds the strength of the overburden.

Morgan, J.P. 1961. Genesis and paleontology of the Mississippi River mudlumps; Part 1—Mudlumps at the mouths of Mississippi Rivers. Louisiana Geol. Survey Bull., v. 35, 115 pp.

Mudlumps are topographic highs formed by the intrusion of diapirs or anticlines of fine-grained sediment into rapidly depositing deltaic sediments. They are typically about a hectare in area and are commonly capped by mud volcanoes. They represent a type of very early structural activity that gives rise to juxtaposition of sediment types. This reference contains some 60 photographs. See Hamilton (1976) on similar features in Indonesia.

Morgan, J.P., J.M. Coleman, and S.M. Gagliano. 1968. Mudlumps—Diapiric structures in Mississippi Delta sediments. *In:* J. Braunstein and G.D. O'Brien, Eds., Diapirism and Diapirs, Amer. Assoc. Petrol. Geol. Mem. 8, p. 145–161.

Detailed account and results of shallow test boring program. The authoritative paper.

Oertal, G., and C.D. Curtis. 1972. Clay–ironstone concretions preserving fabrics due to progressive compaction. Geol. Soc. Amer. Bull., v. 83 p. 2597–2606.

Fabrics of basal planes of clay minerals are related to stages in the growth of a clay–ironstone concretion.

Otvos, E.G. 1970. High pressure shales and their deposition facies, southern Louisiana Cenozoic: Jour. Sed. Petrol., v. 40, p. 412–417.

Undercompacted shales, containing more water than normal for their burial depth, give low-resistivity log readings. They increase in abundance basinward.

Powers, M.C. 1967. Fluid release mechanisms in compacting marine mudrocks and their importance in oil exploration: Amer. Assoc. Petrol. Geol. Bull., v. 51, p. 1240–1254.

Suggests that a good oil source rock requires initial deposition of smectite, deep burial, faulting, and migration. His Figs. 1 (Fig. 3.76) and 3 (not shown) are good. Excellent deductive paper. See also Dickey (1972).

Rieke, H.H., and G.V. Chilingarian. 1974. Compaction of Argillaceous Sediments. Elsevier Scientific Publ. Co., Amsterdam, 424 pp.

Scott W. Starratt
Dept. of Paleontology
U. C. Berkeley
Berkeley, Ca. 94720

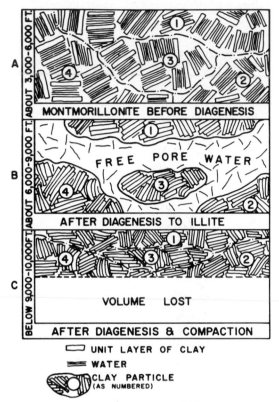

Figure 3.76 Clay diagenesis and compaction of mudrocks. Diagram illustrates how initial "open" structure of montmorillonite (A) first transforms to give free pore water and denser illite (B) and finally a greatly reduced volume (C) when water is expelled and illite is compacted even more. (Powers 1967, Fig. 1; published by permission of the author and the American Association of Petroleum Geologists.)

Comprehensive review of the behavior of fine-grained sediments during compaction. Some treatment of overpressuring and interstitial fluids. Many Russian references.

Rumeau, J.-L., and C. Sourisse. 1972. Compaction, diagenese et migration dans les sediments argileux: Bull. Centre Rech. Pau SNPA, v. 6, p. 313–345.

Fundamental paper relating compaction, diagenesis, and migration, with discussion of both general principles and diverse specific examples. *Key ideas:* Under-and overcompaction, "hydraulic covering," relict compaction, wire-line logs, and factor analyses.

Schmidt, G.W. 1973. Interstitial water composition and geochemistry of deep Gulf Coast shales and sandstones: Amer. Assoc. Petrol. Geol. Bull., v. 57, p. 321–331; Discussion, p. 715–721.

Fluid composition may have a strong influence on shale diagenesis, including compaction and resulting structural effects. One of the few papers to report fluid compositions from both shales and sands. In normally pressured zones, shales have lower salinities than associated sands, whereas in overpressured zones, the waters in the shales have about the same concentration as the sands, both being significantly more dilute than waters in the normal pressure zone.

Sharp, J.M. 1976. Momentum and energy balance equations for compacting sediments: Math. Geol., v. 8, p. 305–322.

One of the latest papers on the problems of explaining the compaction of thick sedimentary masses. A series of partial differential equations is constructed and used to show that excess fluid pressure in the Gulf Coast is caused primarily by compaction disequilibrium.

Shelton, J.W. 1962. Shale compaction in a section of Cretaceous Dakota Sandstone, northwestern North Dakota: Jour. Sed. Petrol., v. 32, p. 873–877.

A simple way to estimate compaction in shales is to unravel penecontemporaneous contorted sandstone dikes. By so doing, a compaction of 260% was calculated—a figure that compares well with other types of estimates. Most probably, cores are needed for this method.

Stel, J.H. 1976. Clay diapirism in the Lower Emsian La Vid Shales near Colle, Cantabrian Mountains, N.W. Spain: Geol. Mijnbouw, v. 55, p. 110–116.

Soft-sediment deformation of clays has produced slumplike structures involving more rigid, carbonate layers, but these are unrelated to the paleoslope and are caused simply by the dewatering process.

White, W.A. 1961. Colloidal phenomena in sedimentation of argillaceous rocks: Jour. Sed. Petrol., v. 31, p. 560–570.

Laboratory experiments and field observations suggest that unflocculated clays have fewer slickensides, syneresis cracks, fissures, pits and mounds, and cones and craters than do those that have been flocculated. Thus, syneresis cracks in shales and mudstones do not necessarily indicate subaerial exposure—but they probably do when they are very abundant.

Diagenesis

There is a large literature on shale diagenesis, because clay mineral transformations are easily studied by X-ray diffraction combined with chemical analysis. Some attention has also recently been devoted to the chemistry of the associated waters. This water chemistry has a possible effect on compaction and, therefore, some references are also included under that heading.

Early diagenesis seems to have very little effect on clays, but an important later reaction is the conversion of smectite to illite, which is now fairly well understood. Concretions still intrigue geologists, but their study has not contributed as much to our geologic knowledge as it probably can. Do they retain information on fabric or depositional environment or provenance now lost from the enclosing rock?

Important advances have also been made in our understanding of the diagenesis of shale organic matter. There is a large, scattered literature on this subject. We have included a few of the most recent, most general references, from which you should be able to find many more.

Aronson, J.L., and J. Hower. 1976. Mechanism of burial metamorphism of argillaceous sediment: 2 Radiogenic argon evidence: Geol. Soc. Amer. Bull., v. 87, p. 738–744.

> Whole-rock apparent K–Ar ages decrease with depth in Gulf Coast shaly sediments, because Ar is lost from K-mica and K-feldspar during their destruction to yield the K used in converting smectite to illite, thus conforming the isochemical nature of this diagenetic transformation.

Björlykke, K. 1973. Origin of limestone nodules in the Lower Paleozoic of the Oslo region: Norsk Geol. Tidsskift, v. 53, p. 419–431.

> Limestone nodules, some up to 2 m in diameter, in both the black alum and gray shales were studied, mostly by careful field observations (with very good pictures) and numerous chemical analyses. Notes all transitions between continuous beds and isolated nodules. Believes many nodules result from dissolution caused by undersaturated sea water. Good model paper.

Blatt, H., and B. Sutherland. 1969. Intrastratal solution of nonopaque heavy minerals in shales: Jour. Sed. Petrol., v. 39, p. 591–600.

> To our knowledge one of few heavy mineral studies of shales. Preservation of heavy minerals was found to be better in shales than in associated sands, so shales are in that sense more suitable for provenance studies than sands. Advanced diagenesis would probably affect heavy minerals in shales more, however, because of higher surface area. A study of a long vertical profile in a shale sequence by this technique would be very interesting.

Claypool, G.E., A.H. Love, and E.K. Maughan. 1978. Organic geochemistry, incipient metamorphism, and oil generation in black shale members of Phosphoria Formation, western interior United States: Amer. Assoc. Petrol. Geol. Bull., v. 62, p. 98–120.

> Organic maturity, as indicated by amount and type of extractables and by kerogen color, shows that most oil has been generated in these rocks at temperatures corresponding to burial depths of 2.5–4.5 km. Has an interesting isopach of total carbon plus a depth versus hydrocarbon generation plot (Fig. 3.77).

Drever, J.I. 1971. Early diagenesis of clay minerals, Rio Ameca Basin, Mexico: Jour. Sed. Petrol., v. 41, p. 982–994.

> In oxidizing sediments, there is no early diagenetic clay mineral reaction other than ion exchange, but in reducing sediments, there is some additional uptake of Mg^{2+}, presumably caused by exchange for Fe^{2+} produced by the reduction of Fe^{3+} contained in the clays.

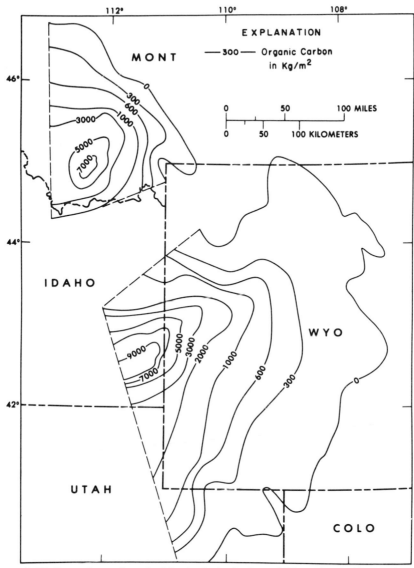

Figure 3.77 Organic carbon in black shale facies of Phosphoria Formation (kg/sq m). (Claypool and others 1978, Fig. 9; published by permission of the authors and the American Association of Petroleum Geologists.)

Dunoyer de Segonzac, G. 1970. The transformation of clay minerals during diagenesis and low grade metamorphism: A review: Sedimentology, v. 15, p. 281–346.

A long review of the changes in clay minerals during burial, which is far the best introduction to this subject. Reactions involving smectite and illite are fairly well understood, but much more work on chlorite is needed, such as the paper by Hayes (1970).

Durand, B., J. Espitalié, G. Nicaise, and A. Combaz. 1972. Étude de la Matière Organique Insoluble (Kérogene) des Argiles du Toarcien du Bassin de Paris; Premiere Partie, Étude par les Procedes Optiques, Analyse Elementaire, Étude en Microscopie et Diffraction Electroniques: Rev. Inst. Francais du Pétrole, v. 27, p. 865–884.

Optical, electron microscopy, and electron diffraction analyses show that the properties of kerogen change gradually in burial. (English and Spanish abstracts.)

Etheridge, M.A., and M.F. Lee. 1975. Microstructure of slate from Lady Loretta, Queensland, Australia: Geol. Soc. Amer. Bull., v. 86, p. 13–22.

Micas in this dolomitic slate show a bimodal distribution of orientations. Larger grains are predominantly aligned parallel to bedding, whereas smaller grains are parallel to cleavage, suggesting a solution-precipitation mechanism for development of the cleavage. See Holeywell and Tullis (1975).

Gilman, R.A., and W.J. Metzger. 1967. Cone-in-cone concretions from western New York: Jour. Sed. Petrol., v. 37, p. 87–95.

This common sedimentary structure of black shales is said to be caused by volume increase when primary aragonite inverts to calcite.

Hallam, A. 1962. A band of extraordinary calcareous concretions in the Upper Lias of Yorkshire, England: Jour. Sed. Petrol., v. 32, p. 840–847.

Concretions of both calcisiltstone and calcilutite in finely laminated bituminous shales and marls. This paper explores their diagenetic history.

Hiltabrand, R.R., R.E. Ferrell, and G.K. Billings, 1973. Experimental diagenesis of Gulf Coast argillaceous sediment: Amer. Assoc. Petrol. Geol. Bull., v. 57, p. 338–348.

Experimental verification of field studies, such as Perry and Hower (1970) and Weaver and Beck (1971) shows that the observed clay mineral diagenesis can occur in a closed system without addition of new ions. Table 2 (not shown) summarizes the observed diagenetic sequences reported in five field studies.

Holeywell, R.C., and T.E. Tullis. 1975. Mineral reorientation and slaty cleavage in the Martinsburg Formation, Lehigh Gap, Pennsylvania: Geol. Soc. Amer. Bull., v. 86, p. 1296–1304.

Cleavage in these rocks appears to be produced by recrystallization of mica and chlorite rather than by mechanical rotation. The X-ray technique used to measure orientation would be excellent for fabric analysis in other shale-containing units. See Fig. 3.78.

Johns, W.D., and A. Shimoyama. 1972. Clay minerals and petroleum-forming reactions during burial and diagenesis: Amer. Assoc. Petrol. Geol. Bull., v. 56, p. 2160–2167.

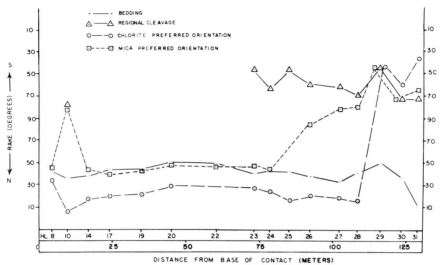

DISTANCE FROM BASE OF CONTACT (METERS)

Figure 3.78 Orientation of bedding, regional cleavage, mica, and chlorite as a function of distance below overlying conglomerate. Near the contact most of the micas and chlorites lie in the bedding plane but away from it, first the micas and then the chlorites diverge and align with the regional cleavage. This shows that the cleavage is the result of recrystallization rather than mechanical reorientation. (Holeywell and Tullis 1975, Fig. 7; published by permission of the authors and the Geological Society of America.)

Key ideas: Burial depth, water expulsion, alkane production (decarboxylation), maturation (cracking), diagenesis, and experiments.

Long. G., S. Neglia, and L. Favretto. 1968. The metamorphism of the kerogen from Triassic black shales, southeast Sicily: Geochim. Cosmochim. Acta, v. 32, p. 647–656.

In these rocks, diagenesis is reported to result in a decrease in the H/C ratio, with O/C remaining fairly stable; an increase in aromatic relative to aliphatic hydrocarbons; and a graphitization of the carbon as indicated by increasing sharpness of the 002 graphite peak. These changes have apparently led to the release of abundant methane from the rocks.

Lucas, J., and G. Ataman. 1968. Mineralogical and geochemical study of clay mineral transformations in the sedimentary Triassic Jura Basin (France): Clays Clay Minerals, v. 16, p. 365–372.

Maintains that the basinward sequence of poorly crystallized illite through mixed-layer illite–chlorite is caused by syndepositional reactions, not by diagenesis, and that clays do rapidly adjust to their environment of deposition. Compare with Dunoyer de Segonzac (1970).

Massaad, M. 1973. Les Concrétions de ''l'Aalenien.'': Schweizerische Mineral. Petrograph. Mitt., v. 53, p. 405–459.

A very comprehensive study emphasizing concretions as overlooked keys to the origins of their hosts (especially shales) and as guides to burial depth, paleogeography, metamorphism, physiochemical conditions of sedimentation, and possibly even paleoclimates. Many chemical analyses, plus a general classification of concretions and 10 conclusions.

Maxwell, D.T., and J. Hower. 1967. High-grade diagenesis and low-grade metamorphism of illite in the Precambrian Belt series: Amer. Mineral., v. 52, p. 843–857.

During burial, the illites in this rock sequence were transformed from the 1 Md to the 2 M polymorph. This change occurs gradually over an appreciable stratigraphic thickness and therefore would be a useful indicator of burial history.

McNamara, M. 1965. The Lower Greenschist Facies in the Scottish Highlands: Geol. fören Stockholm, v. 87, p. 347–389.

Mostly focused on mineralogy. For example, "the clay minerals are replaced by 2 M muscovite or phengite and 14 Å chlorite," but there is a little (pp. 348–349) discussion of bedding. Primarily. Turner-type metamorphic petrology.

• Mitsui, K., and K. Taguchi. 1977. Silica mineral diagenesis in Neogene Tertiary Shales in the Tempoku District, Hokkaido, Japan: Jour. Sed. Petrol., v. 47, p. 158–167.

Diatomaceous marine shales show a diagenetic sequence of SiO_2–opal cristobalite-low cristobalite–low quartz. The intensity of the (101) quartz reflection and the crystallinity index follow these changes.

Perry, E.A. Jr. 1974. Diagenesis and the K–Ar dating of shales and clay minerals: Geol. Soc. Amer. Bull., v. 85, p. 827–830.

Measured K–Ar ages of mixed-layer illite/smectite become younger with depth, because of diagenetic addition of potassium to interstratified smectite–illite layers (Fig. 3.79). See also Powers (1967), Dickey (1972), and Aronson and Hower (1976).

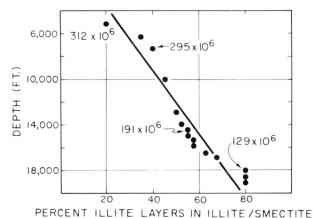

Figure 3.79 Change of "age of illite" with depth in the Gulf Coast. (Redrawn from Perry 1974, Fig. 1).

Perry, E.A. Jr., and J. Hower. 1970. Burial diagenesis in Gulf Coast Pelitic sediments: Clays Clay Minerals, v. 18, p. 165–177.

> Vertical profiles from deep wells in the Gulf Coast show that the major diagenetic change in clay minerals is the conversion of mixed-layer illite/smectite to illite accompanied by incorporation of K^+. A major landmark in the diagenetic, burial history of clay minerals, comparable to studies of clay mineral distribution in the modern muds of the world ocean by Griffin, and others. (1968) and Rateev, and others. (1969).

Russell, K.L. 1970. Geochemistry and halmyrolysis of clay minerals, Rio Ameca, Mexico. Geochim. Cosmochim. Acta, v. 34, p. 893–907.

> Shows that the only important reaction between river-borne clays and sea water is ion exchange. Contains valuable analyses of exchangeable and non-exchangeable cations.

Scholle, P.A., and P.R. Schluger, Eds. 1979. Aspects of Diagenesis: Soc.Econ. Paleont. Mineral. Sp. Pub. 26, 443 pp.

> Part 1, Determination of diagenetic paleotemperature, contains 5 papers which are especially relevant to shales.

Tullis, T.E. 1976. Experiments on the origin of slaty cleavage and schistosity: Geol. Soc. Amer. Bull., v. 87, p. 745–753.

> Experimental evidence suggests that mineral reorientation, producing cleavage or schistosity, is facilitated by recrystallization of the grains, which reduces stresses along their boundaries. The grains therefore behave much as though suspended in a homogeneous, deforming fluid. High fluid pressures may similarly "catalyze" this reorientation.

van Moort, J.E. 1971. A comparative study of the diagenetic alteration of clay minerals in Mesozoic shales from Papua, New Guinea, and in Tertiary shales from Louisiana, U.S.A.: Clays Clay Minerals, v. 19, p. 1–20.

> The smectite to illite transformation observed in Gulf Coast shales (cf. Perry and Hower 1970, Weaver and Beck 1971) also occurs in shales derived from a different type of source area in a different climate.

Weaver, C.E. 1967. Potassium, illite and the ocean: Geochim. Cosmochim. Acta, v. 31, p. 2181–2196.

> A diagenesis study of potassium in muds and shales with a notable estimate of clay mineral composition in North America versus time. Please see his redrawn Fig. 3.80.

Weaver, C.E., and K.C. Beck. 1971. Clay–Water Diagenesis During Burial: How Mud Becomes Gneiss. Geol. Soc. Amer. Spec. Paper 134, 96 pp.

> Detailed description of the chemical and mineralogic changes in shales during diagenesis. They suggest that some K^+ and Mg^+ must be added to the rocks by upward migrating fluids in order for the later diagenetic changes to occur. Compare with Aronson and Hower (1976) and Hiltabrand and others (1973).

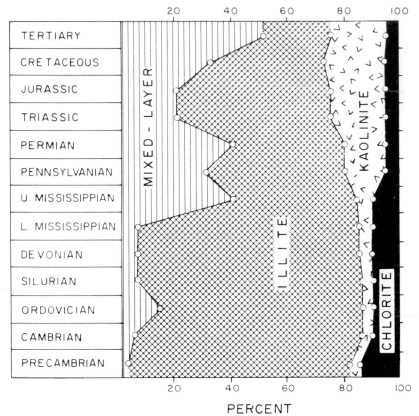

Figure 3.80 Change in proportions of major clay mineral groups based on over 40,000 analyses from North America. (Redrawn from Weaver 1967, Fig. 3.)

Thermal History

One of the most important components of a basin analysis is defining its thermal history and this can often be done best by studying its shales. Great advances in this area are being made, some of the more successful indicators being illite crystallinity, conodont color, and vitrinite reflectance.

Bostick, N.H., and V.N. Foster. 1973. Comparison of vitrinite reflectance in coal seams and in kerogen of sandstones, shales and limestones in the same part of a sedimentary section. *In:* B. Alpern, Ed. Collogue International Petrographie de la Matiero Organique des Sédimentes, Centre National de la Recherche Scientifique, Paris, p. 13–25.

The reflectance of coaly particles in sedimentary rocks, including shales, is an important indicator of thermal history. This paper shows reflectance values to also be a function of lithology.

Bostick, N.H., and C.P. Nicksic. 1975. Bibliography and Index of Coal and Dispersed Organic Matter in Sedimentary Rocks; Petrography, Catagenesis, Relation to Petroleum and Natural Gas, and Geochemistry. Illinois State Geol. Survey, Illinois Petrol. No. 108, 92 pp.

Good source when looking for more information about organic matter in shales, especially for European references.

Epstein, A.G., J.B. Epstein, and L.D. Harris. 1977. Conodont Color Alteration: An Index to Organic Metamorphism. U.S. Geol. Survey Prof. Paper 995, 27 pp.

Rarely published colored plates effectively tell the story of how conodont color changes with paleotemperatures.

Foscolos, A.E., F.G. Powell, and P.R. Gunther. 1976. The use of clay minerals and inorganic geochemical indicators for evaluating the degree of diagenesis and oil generating potential of shales: Geochim. Cosmochim. Acta, v. 40, p. 953–966.

A valuable study because it combines the various temperature indicators for sedimentary rocks (vitrinite reflectance, mixed layering in clays, illite crystallinity and polymorphism, C/H/O ratios, hydrocarbon analyses) on the same vertical sequence. Also includes careful mineralogic and chemical analyses. Unfortunately, a new set of Greek names is presented for various stages of diagenesis.

Grew, E.S. 1974. Carbonaceous material in some metamorphic rocks of New England and other areas: Jour. Geol., v. 82, p. 50–73.

Traces the changes in organic matter with increasing metamorphic grade from chlorite to sillimanite grades. Carbon increases at the expense of H, N, O until a pure carbon phase is found in staurolite-grade rocks. The peak width and 2θ position of the 002 graphite reflection give a measure of the thermal grade.

Gutjahr, C.C.M. 1966. Carbonization measurements of pollen grains and spores and their application: Leidse Geol. Med., v. 38, p. 1–29.

Describes a method for estimating paleotemperature based on the amount of light absorbed by pollen grains and gives examples of its application to two Gulf Coast areas.

Hoefs, J., and M. Frey. 1976. The isotopic composition of carbonaceous matter in a metamorphic profile from the Swiss Alps: Geochim. Cosmochim. Acta, v. 40, p. 945–951.

The metamorphism of a black shale is described and related to carbon isotopes. δC^{13} values are stable at about -25 per mil up to greenschist grade but then steadily increase to -11 per mil in the amphilbolite grade rocks.

Hood, A., and J.R. Castaño. 1974. Organic metamorphism: Its relationship to petroleum generation and application to studies of authigenic minerals: United Nations Econ. Comm. Asia Far East Coord. Comm. Offshore Prospecting Techology Tech. Bull., v. 8, p. 85–118.

Review of organic metamorphism describing techniques for measuring its degree (discussing the different hydrocarbon products to be expected) and the relationship of these organic temperature scales to the occurrence of zeolites. One hundred references to the worldwide literature on these subjects.

Hood, A., C.C.M. Gutjahr, and R.L. Heacock. 1975. Organic metamorphism and the generation of petroleum: Amer. Assoc. Petrol. Geol. Bull., v. 59, p. 986–996.

Defines a scale of organic matter alteration. This scale is compared to coal rank, palynormorph color, and vitrinite reflectance. Time versus temperature problems are emphasized, and the activation energy for the various changes is calculated. Forty-five references.

Hosterman, J.W., G.H. Wood, and M.J. Bergin. 1970. Mineralogy of Underclays in the Pennsylvania Anthracite Region. U.S. Geol. Survey Prof. Paper 700-C, p. C89–C97.

Describes metamorphic alteration of underclays to greenschist facies. Kaolinite, 2M illite, chlorite, pyrophyllite, and phlogopite are present.

Kisch, H.J. 1974. Anthracite and meta-anthracite coal ranks associated with "anchimetamorphism," "very low stage" metamorphism. I, II, and III: Koninkl. Nederl. Akad. Wetenschappen Amsterdam Proc. Ser. B, v. 77, p. 81–118.

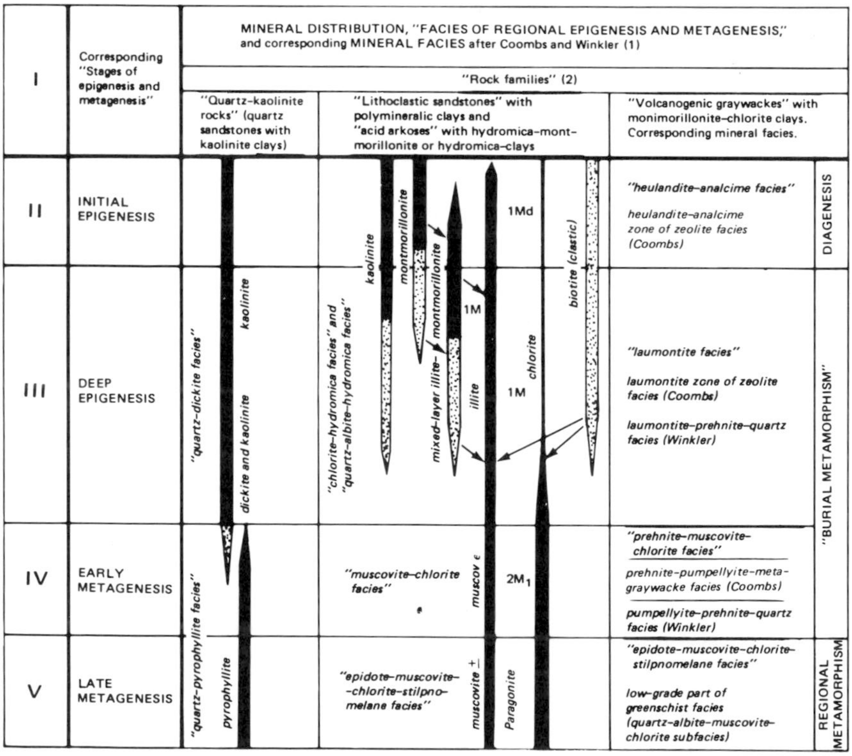

Figure 3.81 Stages of metamorphism and mineral composition from diverse authors. (Kisch 1974, p. 85.)

This three-part paper sums up the results of a long interest by the author in low-rank metamorphism of coals and their associated shales and sandstones. Useful tables and figures correlate the responses of clay minerals and coals to the indicator minerals used by metamorphic petrologists. One hundred and six references. See Fig. 3.81.

Shibaoka, M., A.J.R. Bennett, and K.W. Gould. 1973. Diagenesis of organic matter and occurrence of hydrocarbons in some Australian sedimentary basins: Jour. Australian Petrol. Explor. Assoc., p. 73–80.

Discussion of the petroleum potential of several Australian depositional basins, using vitrinite reflectance to define the "oil window." In some basins erosion has been deep enough to completely remove this oil zone.

Snowden, L.R., and K.J. Roy. 1975. Regional organic metamorphism in the Mesozoic strata of the Sverdrup Basin: Canadian Petrol. Geol. Bull., v. 23, p. 131–148.

Application of maturation measures to a basin analysis in the Arctic Islands. Illustrations show the distribution of immature, mature, and metamorphosed hydrocarbon facies for three different formations.

Staplin, F.L. 1969. Sedimentary organic matter, organic metamorphism, and oil and gas occurrence: Canadian Petrol. Geol. Bull., v. 17, p. 47–66.

Uses type and color of organic matter to delineate areas favorable for the accumulation of dry gas, wet gas, and petroleum. Nine examples, in color, of this technique.

SHALE THROUGH TIME

A continuing question in the study of shales is whether their volume and composition has changed with time. Such changes might reflect changes in the earth's crustal composition or in the chemistry of the oceans.

Garrels, R.M., and F.T. MacKenzie. 1971. Evolution of Sedimentary Rocks. Norton and Co., New York, 397 pp.

Chapter 9 contains a discussion of trends in shale composition with time. In general, older shales are enriched in K_2O and depleted in Na_2O and CaO. These changes can be explained by increased diagenesis in the older rocks (e.g., conversion of smectite to illite).

Leventhal, J., S.E. Suess, and P. Cloud. 1975. Nonprevalence of biochemical fossils in kerogen from pre-Phanerozoic sediments: Proc. Natl. Acad. Sci., Washington, D.C., v. 72, p. 4706–4710.

Biochemical fossils are unlikely to be found in Precambrian rocks because of extensive alteration of the organic matter. Gives 14 references to Precambrian shale occurrences.

Nanz, R.H. 1953. Chemical composition of Precambrian slates with notes on the geochemical evolution of lutites: Jour. Geol., v. 61, p. 51–64.

One of the first papers to examine chemical variation of lutites with time and the possible effects of organic control, texture, and source materials.

Ronov, A.B., and A.A. Migdisov. 1971. Geochemical history of the crystalline basement and the sedimentary cover of the Russian and North American platforms: Sedimentology, v. 16, p. 137–185.

Discusses evolution of composition of crustal rocks, with a number of diagrams for argillaceous sediments. In general, iron is more oxidized in younger sediments, Fe/Mn is higher, and organic carbon increases sharply. There is a pronounced maximum in K_2O in the early Paleozoic.

van Moort, J.C. 1972. The K_2O, CaO, MgO and CO_2 contents of shales and related rocks and their importance for sedimentary evolution since the Proterozoic. Internatl. Geol. Cong., 24th Sess., Sec. 10, ed. J. Gill, Montreal, Canada, p. 427–439.

Shales from the Australian platform show increasing K_2O and increasing MgO/CaO with increasing age, in agreement with other areas. This paper suggests that these changes are secondary and result from transfer of MgO and K_2O from associated sandstones and volcanics, CaO being lost from the system.

ECONOMIC GEOLOGY

We have selected a few papers, usually reviews, that cover various economic aspects of shales. Shale-associated mineral deposits are primarily the shale itself, as either clays or lightweight aggregate, or sulfide minerals associated with the shale. In addition, we have included references on "bituminous" shale, that sometimes enigmatic rock type that in some basins is an important source of natural gas and holds great potential for future petroleum resources. It is interesting that the sedimentology of both commercial clay deposits and marine black shales is not very well understood. Perhaps some of the papers cited in earlier sections of this chapter can be applied with profit to improve our understanding of both.

Petroleum and Bituminous Shales

Organic-rich shales, long a source of wonder for sedimentologists and long viewed as a *potential fuel source,* are once again back in the forefront of economic interest. The two cited bibliographies provide many additional references.

Bigarella, J.J. 1972. Geologia da Formacão Irati. In: Simposio sobre Ciencia e Technologia do Xisto, Curitiba. Academia Brasileira de Ciencias, Rio de Janiero, 81 pp.

Careful complete description and interpretation of a famous Brazilian black oil shale. See Fig. 3.82.

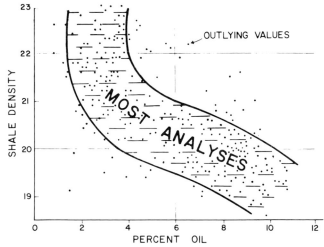

Figure 3.82 Scatter diagram of shale density and percent oil. (Redrawn from Bigarella 1972, Fig. 25.)

Bitterli, P. 1963. Classification of bituminous rocks of western Europe. Sec. I–Geology and geophysics. *In:* Proc. 6th World Petroleum Conf., Frankfurt/Main, p. 155–165.

Really a summary of the different ways of looking at an argillaceous sediment—palynologically, petrographically, and geochemically—and how to classify one based on systematic study of over 1,500 samples. Figure 3.83 is a genetic flow chart, which appears to be a useful general model.

Board on Mineral Resources. 1976. Natural Gas from Unconventional Geologic Sources. Natl. Acad. Sci.–Natl. Res. Council, Commission on Natl. Resources, Washington, D.C., 245 pp.

Section II, "Gas from Brown Shales," has four articles on the potential and problems of extracting gas from shale, and Section I, "Gas from Geopressured Zones," discusses how overpressured shales contribute gas. Excellent source for up-to-date references on a currently very important topic.

Borrello, A.V. 1956. Antecedentes relativos al conocimiento de los esquistos y rocas bituminosas del pais. *In:* Recursos Minerales de la República Argentina, Pt. III, Combutibles Sólidos Minerales. Revista Musco Argentino de Ciencias Naturales "Bernardino Rivadavia" (Buenos Aires), Ciencias Geologicas, v. 5, pp. 581–642.

The chapter on oil shales (esquistos bituminosos) is very complete with 145 references, some from the middle of the nineteenth century. After a brief section on the origin of bituminous shales, individual occurrences are summarized and then described in detail. Argentine oil shales range in age from Carboniferous to Neogene.

Bradley, W.H. 1970. Green River oil shale—Concept of origin extended; An inter-

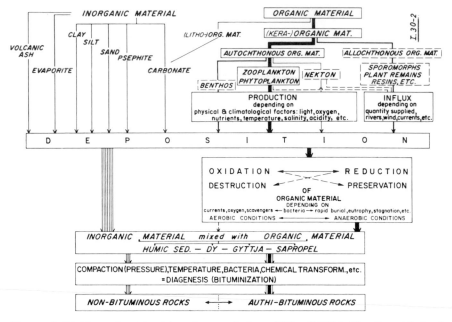

Figure 3.83 Flow chart for bituminous shales. (Bitterli 1963, Fig. 2; published by permission of the author and the Koninklijk Nederl. Geologisch Mijnboukundig Genootschap.)

disciplinary problem being attacked from both ends: Geol. Soc. Amer. Bull., v. 81, p. 985–1000.

A comprehensive review by a lifelong student of the Green River oil shale stresses the need for integrated study by geologists, paleontologists, organic geochemists, biogeochemists, and limnologists. Topics covered include geology, organic chemistry, and a modern-day analog with much emphasis on microscopic algae that accumulated as oozes far from the shoreline of this 50 million-year-old lake, one that now contains over 160 billion barrels of oil in shale with a yield of 30–35 gallons per ton. Over 80 references.

Brongersma-Sanders, M. 1971. Origin of major cyclicity of evaporites and bituminous rocks: An actualistic model: Mar. Geol., v. 11, p. 123–144.

Relates the common association of black shales and evaporites to a combination of geography and climate which can also be used to explain cyclicity in such sequences. Can the Kupferschiefer–Zechstein be explained this way?

Conant, L.C., and V.E. Swanson. 1961. Chattanooga Shale and Related Rocks of Central Tennessee and Nearby Areas. U.S. Geol. Survey Prof. Paper 357, 91 pp.

Mostly careful internal stratigraphy but contains some maps of sandstone thickness within the shale and cross sections that reveal systematic internal facies variations. There are only a few studies that examine the internal stratigraphy of a widespread shale (cf., however, Davis 1970). See Fig. 3.84.

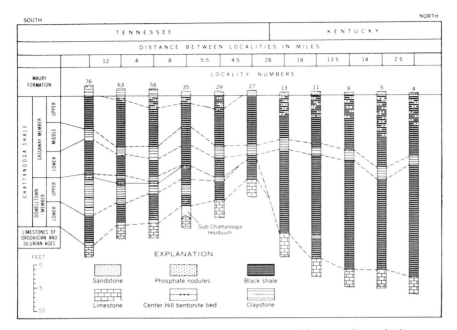

Figure 3.84 Internal stratigraphy of the thin, widespread, and the very famous Chattanooga Shale, where it is very thin. (Conant and Swanson 1961, Fig. 11.)

Duncan, D.C., and V.E. Swanson. 1965. Organic-Rich Shale of the United States and World Land Areas. U.S. Geol. Survey Circ. 523, 30 pp.

   Review of the distribution of organic matter in bituminous shales. One hundred and twenty-one references.

Dunham, K. 1961. Black shale, oil and sulphide ore: Advancement Sci., v. 18, p. 1–16.

   Discussion of how organic matter is preserved in black shale, its conversion into hydrocarbons and their concentration of heavy metals. Sections include "Black Mud," "Bacterial Composition," "Black Shale," "Carbon Compounds and Petroleum Chemistry," and "Sulphide Ores." Well-written overview with 66 references.

Goldhaber, Martin. 1978. Euxinic Facies. In The Encyclopedia of Sedimentology, ed. by R.W. Fairbridge and J. Bourgeois, Eds.: Dowden, Hutchinson and Ross, Inc., Stroudsburg, Pa., p. 296-300.

   Short informative overview with 17 references.

Hard, E.W. 1931. Black shale deposition in central New York: Amer. Assoc. Petrol. Geol. Bull., v. 15, p. 165–181.

   An analysis of depositional conditions of the Devonian black shale—believed to be of shallow-water origin—begins with the sentence "Eventually the oil

EVOLUÇÃO DA MATÉRIA ORGÂNICA
FOLHELHO RADIOATIVO DO MEMBRO BARREIRINHA

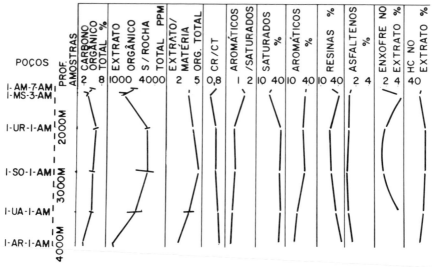

Figure 3.86 Vertical variation of chemical composition of a black shale. (Rodriques 1973, Fig. 7.)

A landmark regional study using indices of diagenesis—illite crystallinity, organic carbon, total soluble organics, electron spin resonance, shale density, and sandstone porosity—to evaluate the petroleum potential of sediments of a long time span and over a wide area. Numerous maps and much tabled data, plus many references. Outstanding.

Smith, J.W., and K.E. Stanfield. 1965. Oil shales of the Green River Formation in Wyoming. *In:* R.H. DeVoto, and R.K. Bitter, Eds.; Sedimentation of Late Cretaceous and Tertiary Outcrops, Rock Springs Uplift, Wyoming. Wyo. Geol. Assoc. 19th ann. Field Conf., p. 167–170.

Brief account of one of the world's largest deposits of hydrocarbons with emphasis upon sampling and evaluation. Fifteen references.

Smith, J.W., and N.B. Young. 1967. Organic composition of Kentucky's New Albany Shale: Determination and uses: Chem. Geol., v. 2, p. 157–170.

Careful elemental analyses of three core samples shows the C/H ratio of organic matter (11.1) to be higher than the Green River Shales (7.8), but lower than coals (14.0). Valuable because it establishes standard values for comparison. Composite samples were used, and they show a very small westward increase in percent organic carbon in the whole rock that may have some paleodispersal significance. Studies of specific stratigraphic intervals are now needed.

Smith, S.E., and W.K. Overbey Jr., Eds. 1975. Bibliography of Upper Devonian Shale Sequence. U.S. Energy Res. and Develop. Admin. (ERDA), Morgantown Energy Res. Center, Morgantown, West Virginia, 78 pp.

Focused on the Upper Devonian black shale sequence and its associated sediments in the Appalachian Basin. References to stratigraphic names, plus relevant economic and engineering aspects—mostly concerning oil and gas.

Stevenson, D.L., and D.R. Dickerson. 1969. Organic Geochemistry of the New Albany Shale in Illinois. Illinois Geol. Survey Petroleum Paper 90, 11 pp.

A study of the extractable organic matter and total organic carbon of the Devonian black shale sequence in Illinois. Extractables were linearly related to organic carbon, but a contour map of the organic content of the bottom 50 ft. shows great areal variability. A rare "stratigraphic geochemistry" paper.

Trudell, L.G., T.N. Beard, and J.W. Smith. 1970. Green River Formation Lithology and Oil Shale Correlations in the Piceance Creek Basin, Colorado. U.S. Dept. Interior, Rept. Invest 7357, 212 pp.

A geologic framework based on nine bore holes and seven outcrops. Includes cross sections with wire-line logs and lithology, plus a very substantial appendix with description of the sections and 12 plate-size illustrations. Twelve references.

Twenhofel, W.H. 1939. Environments of origin of black shales: Amer. Assoc. Petrol. Geol. v. 23, p. 1178-1198.

A long discussion of where black muds presently are accumulating with a handy summary on the final page. From this, Twenhofel suggests that each black shale deserves its own interpretation—a suggestion with which we sympathsize.

Yen, T.F., and G.V. Chilingar, Eds. 1976. Oil Shale. Elsevier Scientific Publ. Co., Amsterdam, 292 pp.

Fifteen contributors and 12 chapters cover most aspects, from geology to retorting and environmental aspects. Well indexed.

Yen, T.F., Ed. 1976. Science and Technology of Oil Shale. Ann Arbor Science Pub. Inc., Ann Arbor, Michigan, 226 pp.

Fifteen articles, mostly on extraction with one article on oil shales in the United States; 15 colored plates.

Shales as Industrial Minerals

We provide only a few papers here, but many additional ones can be found in the reviews by Cooper (1970) and Patterson and Murray (1975). How many economic geologists and sedimentologists study clays and shales as sources for industrial minerals?

Brownell, W.E. 1976. Structural Clay Products. Springer-Verlag, Wien, New York, 231 pp.

Twelve chapters, mostly directed at structural products, and 120 figures, some of which are colored.

Cooper, J.D. 1970. Clays. *In:* Bur. Mines Mineral Facts and Problems, Eds. staff. U.S. Dept. Interior, Bur. Mines Bull. 650, p. 923–938.

Standard economic treatment with description of industry pattern, supply and projected demand, plus discussion of technology. Thirty-seven references.

Crockett, R.N., Compiler. 1975. Slate. Institute Geol. Sci., Mineral Res. Consult. Comm., Great Britain, Mineral Dossier No. 12, 26 pp.

An industrial mineral note organized into 13 sections, including definition, origin, and nature of slate, and concluding with an industrial appraisal.

Grim, R.E. 1962. Applied Clay Mineralogy. McGraw-Hill Book Co., Inc., New York, 422 pp.

Eight chapters, six of which discuss different uses of clays. Classic.

Grim, Ralph E., and Necip Güven, 1978. Bentonites (Developments in Sedimentology, v. 24): Elsevier Scientific Publishing Co., Amsterdam-Oxford-New York, 256 pp.

Five well illustrated and well referenced chapters including much on X-ray diffraction. The last chapter seems particularly useful and is entitled. "Properties and Uses of Bentonite."

Keller, W.D. 1968. Flint clay and a flint–clay facies: Clays Clay Minerals, v. 16, p. 113–128.

Suggests that there are four facies in a flint–clay deposit: high-alumina clays on structural highs, passing through kaolinite flint clay, and then illite–kaolinite plastic clay into normal marine shales. An example from the Pennsylvanian of Missouri is shown. See Fig. 3.87.

Knechtel, M.M., and S.H. Patterson. 1962. Bentonite Deposits of the Northern Black Hills District, Wyoming and South Dakota. U.S. Geol. Survey Bull. 1082M, p. 893–1030.

Description of reserves of this clay, which is used in drilling mud, as a foundry sand bond, and for bonding taconite core pellets. The bentonite was formed by alteration of volcanic ash derived from the west, based on the direction of thickening of the beds. A volcanic origin is confirmed by the presence of igneous minerals in the coarse fraction, although some sedimentary material is also usually incorporated in the bentonite.

Murray, H.H. 1976. Clay. *In:* R.W. Hagemeyer, Ed. Paper Coating Pigments, Tech. Assoc. Paper Pulp Industry Monograph, Ser. 38, 4th ed., p. 69–109.

Mineralogy, geology, occurrences, physical and chemical properties of clays, as well as their grades and specifications, ending with a discussion of recurring problems. Ten tables and 53 references. Good source to acquire an appreciation of the general economic significance of clay.

Patterson, S.H. 1974. Fuller's Earth and Other Industrial Mineral Resources of the Meigs–Attapulgus–Quincy District, Georgia and Florida. U.S. Geol. Survey Prof. Paper 828, 45 pp.

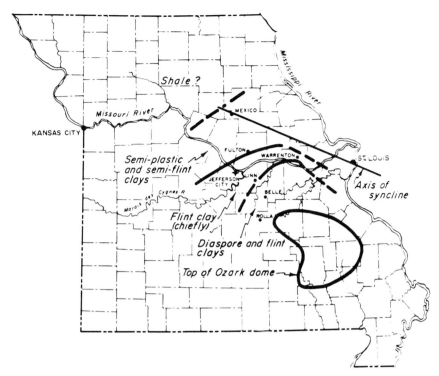

Figure 3.87 Distribution of high-alumina clays, flint clays, plastic clays, and shales of Pennsylvanian age around the Ozark Dome in Missouri. (Keller 1968, Fig. 9; published by permission of the author and the Clay Mineral Society.)

Fuller's earth is fine-grained material of variable mineralogy used as an adsorbent in purification steps in a wide variety of chemical processes. This deposit, which is the leading United States' producer, consists of presumably authigenic palygorskite and sepiolite, and possibly detrital smectite and kaolinite. No generally accepted model has been developed for the origin of the deposits, but they seemed to have formed in very shallow lagoons or tidal flats.

Patterson, S.H., and J.W. Hosterman. 1958. Geology of the clay deposits in the Olive Hill district, Kentucky: Clays Clay Minerals, Proc. 7, Pergamon Press, London, p. 178–194.

Discusses an important kaolinitic flint clay. Interesting in that areal geologic aspects are emphasized. One isopach map and three thin-section microphotographs. See Keller (1968).

Patterson, S.H., and H.H. Murray. 1975. Clays. In: S.J. Lefond, ed.-in-chief. Industrial Minerals and Rocks, 4th ed. New York, p. 518–585.

Definition of terms, history, geology, production, and use, with much, but not total, emphasis upon North America. Includes shales. Over 250 references.

Shales as Hosts for Metallic Minerals

Here you will find one general overview, the paper by Dunham (1961),
plus some papers on the Kupferschiefer of Northern Europe, which is often
considered to be a prototype for many syngenetic ore deposits in shales.

Blissenbach, E., and R. Fellerer. 1973. Continental drift and the origin of certain mineral deposits: Geol. Rundschau, v. 62, p. 812–840.

> Includes a significant section on the properties and formation of metalliferous,
> deep-sea muds formed at early rifting. Also suggests that through transport into
> the subduction zone, metalliferous muds may become a source for porphyry
> ores. Well referenced.

Brongersma-Sanders, M. 1966. Metals of Kupferschiefer supplied by normal sea
water: Geol. Rundschau v. 55, p. 365–375.

> This deposit of bituminous shale is very widespread, at least 20 000 km^2, but is
> only 0.5 m thick and has concentrations of copper, lead, and zinc in the interior
> parts of the basin. Suggests a syngenetic origin by supply of metal ions from nor-
> mal sea water as sluggish surface water moved out at the surface and normal sea
> water moved in at depths. *Key idea:* Upwelling. See also Wedepohl (1964).

Dunham, K. 1971. Introductory talk: Rock association and genesis: Soc. Mining
Geol., Japan, Spec. Issue 3, p. 167–171.

> Brief overview, presented at the 1970 IMA–IAGOD meeting, of the question of
> syngenetic versus epigenetic origin of stratabound sulfide ore bodies, which are
> particularly well developed in shale sequences (e.g., Kupferschiefer, Mount Isa,
> and Zambian Copper Belt), but later metamorphism has often obscured the rela-
> tionships.

Schuchert, C. 1915. The conditions of black shale deposition as illustrated by the
Kupferschiefer and Lias of Germany: Trans. Amer. Phil. Soc., v. 54, p. 259–269.

> An eassay review paper inspired by Pompecj's (1914) "Das Meer des Kup-
> ferschiefers." After an account of this early classic work, Schuchert makes com-
> parisons with the Black Sea, Baltic Sea, and Norwegian Fjords, and concludes
> that "depth of water is not the first essential for the production of foul bottoms"
> (p. 269). See also Brongersma-Sanders (1966).

Smith, G.E. 1974. Depositional Systems, San Angelo Formation (Permian), North
Texas; Facies Control of Red-Bed Copper Mineralization. Texas Bur. Econ. Geol.
Rept. Invest. No. 80, Univ. Texas, 74 pp.

> Commercial grade copper deposits are described from a tidal flat sequence.
> Copper is localized in tidal channel sandstones and in algal mat shales (the
> Flowerpot Mudstone Member). Overlying sabkhalike anhydrite deposits suggest
> mineralization by upward moving ground water accompanied by precipitation
> of copper sulfides by the algal organic matter.

Wedepohl, K.H. 1964. Untersuchungen im Kupferschiefer in Nordwestdeutschland;
Ein Beitrag zur Deutung der Genese bituminöser Sedimente: Geochim. Cosmochim. Acta, v. 28, p. 305–364.

Mostly inorganic geochemistry (trace elements) with much tabled data, one regional map, and many references. Little, however, on facies, thickness, or general geology.

Wedepohl, K.H. 1971. "Kupferscheifer" as a prototype of syngenetic sedimentary ore deposits: Soc. Mining Geol., Japan, Spec. Issue 3, p. 268–273.

The Kupferschiefer is a very unusual and very extensive black shale with high concentrations of copper and other metals which are mined in parts of Germany and Poland. This paper advocates a syngenetic origin of the metals through transgression of a reducing sea over metal-rich red beds. Also makes the point that metallogenesis should also be studied in the subeconomic parts of a deposit.

ENVIRONMENTAL AND ENGINEERING GEOLOGY

A very wide range of papers is included in this section—several source books on soil mechanics, both local and regional studies of engineering properties, clay diapirs, landsliding in argillaceous sediments, soils, and even one reference on landfills! This diverse collection, although only the smallest sampling of a very wide and rapidly growing literature, shows the central role that clays and shales play in the study of surficial processes. We also hope that this collection of papers will be a good comparative source for integrated sedimentologic–engineering studies of enhanced petroleum recovery—especially gas—from shales.

Brookins, D.G. 1976. Shale as a repository for radioactive waste; the evidence from Oklo: Environ. Geol., v. 1, p. 255–259.

Calcareous shale has been an effective seal for a natural nuclear reactor in the Oklo mine of Gabon. Heavy metals have been almost completely trapped, alkaline earth metals have been partly lost, but such elements as iodine seem to have been very mobile.

Bryant, W.R., A.P., Deflache, and P.K. Trabant. 1974. Consolidation of marine clays and carbonates. In: A.I. Inderbitzen, Ed., Deep-Sea Sediments, Plenum Press, New York and London, p. 209–244.

Soil mechanics of marine clays from the Gulf of Mexico reveal a linear relation between void ratio and pressure—in contrast to the situation on land. Also contains discussion of an effect of the role of sedimentation—called delayed consolidation.

Burnett, A.D., and P.G. Fookes. 1974. A regional engineering geological study of the London Clay in the London and Hampshire Basins: Quart. Jour. Eng. Geol., v. 7, p. 257–295.

Authors link regional lithology, paleogeography, and structure of the London Clay to engineering data from site investigations; i.e., they combine regional, qualitative geologic information with local engineering data to form a broad pic-

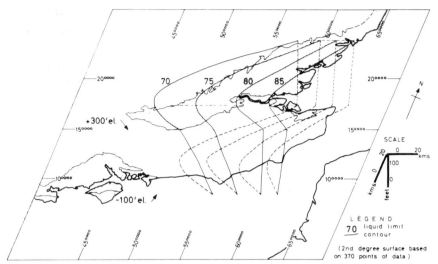

Figure 3.88 Three-dimensional trend surface map of regional distribution of liquid limit. (Burnett and Fookes 1974, Fig. 27; published by permission of the authors and the Geological Society.)

ture. Topics covered include: sedimentology, stratigraphy, mineralogy, lithology, cation-exchange capacity, relation of geologic and soil index properties, and regional trends of engineering index properties using three-dimensional computer trend surface maps. Outstanding and one of a kind. See (Fig. 3.88).

Cabera, J.G., and I.J. Smalley. 1973. Quickclays as products of glacial action: A new approach to their nature, geology, distribution and geotechnical properties: Eng. Geol., v. 7, p. 115–133.

Hypothesizes that quick-clay properties are produced by very fine nonclay particles rather than by the clay minerals. Because glacial grinding produces such minerals in abundance, quick clays are largely confined to glacial deposits. Contrasts strongly with the views of Rosenqvist (1966).

Cartwright, K., and F.B. Sherman. 1969. Evaluating Sanitary Landfill Sites in Illinois. Illinois Geol. Survey, Environ. Geol. Note 27, 15 pp.

Discussion of how clay-rich tills—and presumably shales as well—provide a good impermeable basal seal for the effluent from a landfill.

Crawford, C.B. 1968. Quick-clays of eastern Canada: Eng. Geol., v. 2, p. 239–265.

Good review of the mineralogic and chemical factors producing quick-clay properties. Fifty-two references.

Crooks, J.H.A., and J. Graham. 1972. Stress–strain properties of Belfast estuarine clay: Eng. Geol., v. 6, p. 275–288.

Good example of the types of physical measurements that might be applied to a fine-grained deposit. Unfortunately, no maps of these properties are presented.

Ferguson, H.F. 1967. Valley stress release in the Allegheny Plateau: Eng. Geol., v. 4, p. 63–71.

This paper notes that deep erosional valleys in the Allegheny Plateau have joint systems that are related to valley orientation—possibly because of stress release induced by rapid valley erosion—and surprisingly that this stress field may extend to depths of several thousand feet (p. 68) and so may even be a factor in shallow fractured reservoirs. Shales, unless very plastic, would appear to be very sensitive to such topographic effects.

Franklin, J.A., and R. Chandra. 1972. The slake-durability test: Internatl. Jour. Rock Mech. Min. Sci., v. 9, p. 325–341.

The main purpose of the slake-durability test is to evaluate the weathering resistance of shales, mudstones, siltstones, and related clay-rich rocks. The response varies widely depending upon lithology and stratigraphy.

Galley, J.E. 1968. Subsurface Disposal in Geologic Basins—A Study of Reservoir Strata: Amer. Assoc. Petrol. Geol. Mem. 10, 253 pp.

Deep well disposal is becoming increasingly important for liquid wastes that are hard to treat. This book concentrates on the characteristics of the reservoir zones, but more data are needed on the confining strata, usually shales. If their strength is too low, pumping of waste may lead to hydraulic fracturing of the cap and escape of the waste.

Gillot, J.E. 1968. Clay in Engineering Geology. Elsevier Scientific Publ. Co., Amsterdam, 296 pp.

Twelve short chapters cover a variety of topics, excluding, however, many geologic aspects. Over 700 references and 118 figures.

Gretener, P.E. 1969. Fluid pressure in porous media, its importance in geology—A review: Canadian Petrol. Geol. Bull., v. 17, p. 255–295.

A basic, well-written paper that starts with the concepts of effective stress and failure and gives applications to muds and shales, frozen ground, slope stability, open joints, well fracturing, shale diapirism, and sedimentary structures. Over 80 references.

Hough, B.K. 1957. Basic Soils Engineering. Ronald Press, New York, 513 pp.

An older elementary, but well-recommended, text with 16 chapters and eight appendices.

Hunt, C.B. 1972. Geology of Soils: Their Evolution, Classification, and Uses. W.H. Freeman and Co., San Francisco, 344 pp.

Soils are discussed in 13 chapters from three viewpoints—geology, soil science, and engineering—all in a well-written, easy to read manner. Useful checklist of soil properties.

Krause, H.-F., H.H. Damberger, W.J. Nelson, S.R. Hunt, C.T. Ledvina, C.G. Treworgy, and W.A. White. 1979. Roof Strata of the Herrin [No 6] Coal

Member in Mines of Illinois: Their Geology and Stability. Ill. Geol. Survey, Illinois Minerals Note 72, May, 54 pp.

Very comprehensive study of roof failure in coal mines shows failures to depend upon roof types—does the roof consist of interbedded black shale and limestone or gray shale? Most structural features are believed to have formed early—hence structural trends in the roof and roof lithologies are closely interrelated. Very well illustrated with 48 illustrations and exceptionally well done—a pioneer study to be emulated. See Fig. 3.89

Jackson, J.O., and P.G. Fookes. 1974. The relationship of the estimated former burial depth of the Lower Oxford Clay to some soil properties: Quart. Jour. Eng. Geol., v. 7, p. 137–179.

Engineering properties of this shale were found to vary significantly in different paleontologic zones, but different depths of burial (between 1000 and 5000 ft.) have had little effect. Interesting use of engineering geology on biostratigraphically defined units.

Kendall, H.A. 1974. Clay mineralogy and solutions to clay problems in Norway: Jour. Petrol. Technol., v. 26, p. 25–32.

Drilling problems occur in the Tertiary of the North Sea and most result from Tertiary sedimentation patterns—original, provenance-related differences in clay densities and mineralogy.

Kerr, P.F., and I.M. Drew. 1968. Quick-clay slides in the U.S.A.: Eng. Geol., v. 2, p. 215–238.

Mostly a review of the Anchorage earthquake slides, containing considerable information on the mineralogy and physical properties of the clay deposit.

Lambe, T.W., and R.V. Whitman. 1969. Soil Mechanics. John Wiley and Sons, New York, 553 pp.

Five parts: "Introduction," "The Nature of Soil," "Dry Soil," "Soil with Water—No Flow or Steady Flow," and "Soil with Water—Transient Flow." Well done.

Legget, R.F. 1967. Soil—its geology and use: Geol. Soc. Amer. Bull., v. 78, p. 1433–1459.

Philosophic and historical overview of the role of soils in geology—especially their engineering geology.

Milling, M.E. 1975. Geologic appraisal of foundation conditions, northern North Sea. (Conference Paper) Oceanology Internatl., Brighton, England, March 16–21, BPS Exhibitions Ltd., London, England, p. 311–319.

The engineering properties of different facies of Pleistocene–Holocene sedimentation in the North Sea strongly affect foundation design of drilling platforms. Interesting use of how seismic stratigraphy can be combined with soil mechanics. Excellent. See Fig. 3.90.

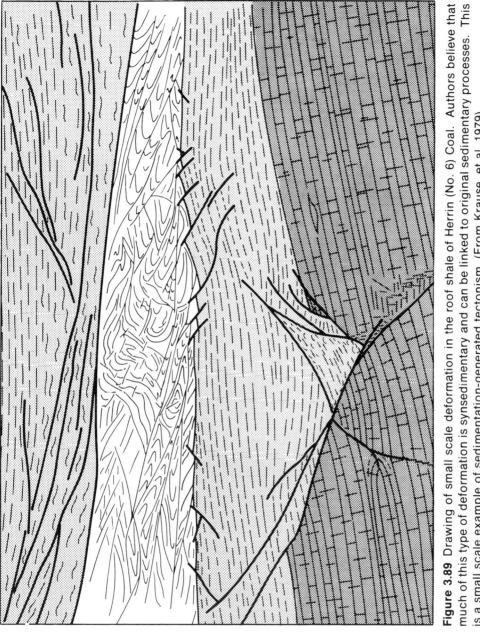

Figure 3.89 Drawing of small scale deformation in the roof shale of Herrin (No. 6) Coal. Authors believe that much of this type of deformation is synsedimentary and can be linked to original sedimentary processes. This is a small scale example of sedimentation-generated tectonism. (From Krause, et al, 1979).

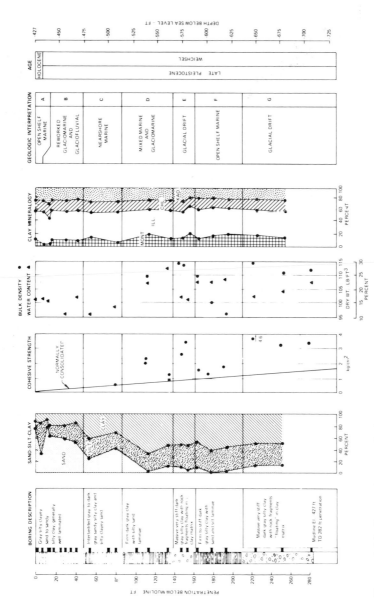

Figure 3.90 Geologic description and laboratory test results of cores recovered from Soil Boring A in offshore Norway study area. (Milling 1975, Fig. 3; published by permission of the author and Oceanology International.)

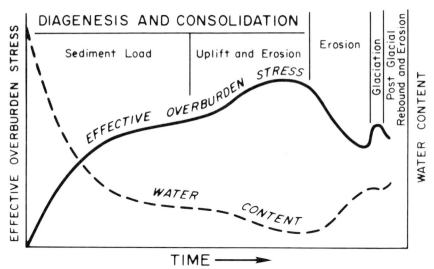

Figure 3.91 History of a shale from deposition to road cut or building foundation. (Redrawn from Scott and Brooker 1966, Fig. 8.)

Moran, S. 1972. Subsurface Geology and Foundation Conditions in Grand Forks, North Dakota. North Dakota Geol. Survey Misc. Ser. 44, 18 pp.

> Texture, clay mineralogy, natural water content, dry density, liquid limit, plasticity index, and unconfined compressive strength for eight units overlying bedrock. Contours top of bedrock and top of five of the units and presents several cross sections. Units 7 and 8 and part of 6 are pre-28000 years B.P.A good, down-to-earth model of what to give the engineering geologist and architect.

Poulet, M. 1976. Apport des expériences de mécanique des roches á la géologie structurale des bassins sédimentaires. Rev. Inst. Francais du Pétrole, v. 31, p. 781–822.

> An essay on the relation between rock strength as measured in the laboratory and the corresponding qualitative behavior in a sedimentary basin as shown in Fig. 3.91.

Rosenqvist, I. Th. 1966. Norwegian research into the properties of quick clay—A review: Eng. Geol., v. 1, p. 445–450.

> Quick clays are a type of fine-grained deposit characterized by a rapid change from an apparently solid to a liquid state, usually upon some shock, such as an earthquake. They may lead to severe landslide damage. This paper summarizes some 20 references to the Norwegian literature dealing with this subject.

Scott, J.S., and E.W. Brooker. 1966. Geological and Engineering Aspects of Upper Cretaceous Shales in Western Canada. Geol. Survey Canada Paper 66-37, 75 pp.

> Relates geologic properties and geologic history of shales to their engineering properties, including the history of those properties and how they affect slope stability. See Fig. 3.92 and their Table 2 (not shown).

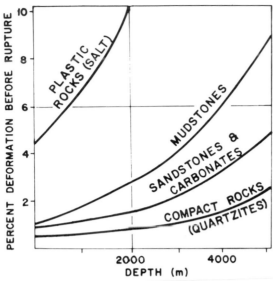

Figure 3.92 Contrasting behavior of major sedimentary rocks prior to failure. (Redrawn from Poulet 1976, Fig. 26.)

Scully, J. 1973. Landslides in the Pierre Shale in Central South Dakota. South Dakota Dept. Transportation, Pierre, S. Dakota, and Fed. Highway Admin., U.S. Dept. Transportation, Washington, D.C., Shale Study 635(67), 707 pp.

> Thirteen chapters plus one of conclusions, with much on the composition of the Pierre Shale, plus its bedding, jointing, ground water, landslides, photogeology, instability, stress deformation, and more.

Smalley, I. 1976. Factors relating to the landslide process in Canadian quickclays: Earth Surf. Process., v. 1, p. 163–172.

> Factors to be considered include nomenclature, size limits, mineralogy, cementation particle coatings, interparticle bond type, soil structure, leaching, and local earthquake intensity and frequency.

Sondhi, V.P., 1966. Note on Landslips on the Dimapur-Manipur Road. *In* Landslides and Hillside Stability in the Eastern Himalayas. India Geol. Surv. Bull., Ser. B, v. 15, Pt. 1, p. 109-117.

> The Dimapur-Manipur Road over the Naga hills in Assam is built on two Tertiary series, which are faulted in the Kohima area. Slips and slumps along the road occur chiefly in the Disarg shales near the fault. Slipping and slumping is caused by rainwater saturation of the broken-up shales in the proximity of the fault, which turns them into a fluid mass. There are two types of movement: the main movement of subsidence and sliding of large areas of the hillside without a complete breaking up of the surface, and a subsequent irregular rolling flow within the loosened material of the slipped mass, which continues after the main slip has taken place. This flow may operate to the end of the rainy season.

Strazer, R.J., L.K. Bestwick, and S.D. Wilson. 1974. Design considerations for deep retained excavations in over-consolidated Seattle clays: Bull. Assoc. Eng. Geol., v. 11, p. 379–397.

Relief of stress by excavation in the Lawton Clay, a glacial lake deposit, produces damaging horizontal expansion. This behavior is apparently caused by the clay's retaining "locked-in" horizontal stress from previous glacial loading. Standard tests tend, therefore, to overestimate the strength of this material.

Terzaghi, K., and R.B. Peck. 1968. Soil Mechanics in Engineering Practice. 2nd ed. John Wiley and Sons, New York, 729 pp.

Advanced reading by a famous author and one of his leading students.

Way, D.S. 1973. Terrain Analysis: A Guide to Site Selection Using Aerial Photographic Interpretation. Dowden, Hutchinson and Ross, Stroudsburg, Pa., 392 pp.

The section on shale (pp.90–97) is really unique in that it describes the appearance of shales on aerial photographs, plus their soil characteristics, topography (in both arid and humid climates), drainage, and typical depths to bedrock as well as all the key issues of site development, of which 10 or more are discussed. Excellent reference for land use.

West, J.M., M.P. Moseley, and D.H. Bennett. 1971. The stability of a valley side in weathered shale: Quart. Jour. Eng. Geol., v. 4, p. 1–23.

Slope stability of shale outcrops is a chronic engineering problem. This study is an example of how such problems can be detected and overcome.

White, W.A., and M.K. Kyriazis. 1968. Effects of Waste Effluents on the Plasticity of Earth Materials. Illinois Geol. Survey Environ. Geol. Note 23, 23 pp.

Expository account that reviews clay mineralogy and relevant engineering terms, such as liquid and plastic limits, and provides tables that show the effects on clays of effluents from septic tanks, oxidation ponds, and landfills.

Wu, T.H. 1966. Soil Mechanics. Allyn and Bacon, Boston, 429 pp.

Thirteen chapters with many illustrations, elementary calculus, and a short chapter on the properties of the clay fraction.

Yong, R.N., and B.P. Warkentin. 1975. Soil Properties and Behaviour. Elsevier Scientific Publ. Co., Amsterdam, 449 pp.

Contains several good chapters of interest in the study of shale, including clay mineralogy, strength measurements, and water movements. Of particular value is the chapter on soil fabric, which discusses its measurement, quantification, and classification.

The real price of anything is the toil and trouble of acquiring it.
Adam Smith

Author Index

Ager, D. 85, 95,126
Akers,W.H. 142
Akhtar, K. 212
Albrecht, P. 189, 191
Allen, J.R.L. 25, 76, 126
Allen, T. 108, 126
Allersma, E. 71, 80, 126, 151, 157, 163,167
Alling, H.L. 19, 126
Amiri-Garroussi, K. 175
Anders, E.R. 119, 126
Anderson, H.V. 202
Andrews, P.B. 212
Angelucci, A., 18,126
Aoyagi, K. 178
Archanquelsky, A.D. 146
Armon, W.J. 262
Armstrong, R.L. 75, 77, 126
Aronson, J.L. 49, 126, 248, 252, 253
Arpino, P. 191
Arthur, M.A. 212
Ashley, G.M. 111, 126
Asmus, H.E. 76, 126
Asquith, D.O. 100, 126, 212, 213, 219
Ataman, G. 251
Atherton, E. 90, 101, 126
Attia, M.I. 161

Bailey, S.W. 127
Baird, G. 100, 121, 126
Baker, D.R. 189
Banks, H.H., Jr. 198
Banner, F.T. 152

Barghoorn, E.S. 138
Barker, C. 59, 126, 241
Baroffio, J.R. 141
Bates, T.F. 193, 200
Beard, J.H. 139
Beard, T.N. 265
Beaudoin, B. 133
Beck, K.C. 144, 250, 253
Beckmann, H. 100, 126
Behrens, E.W. 135
Bennett, A.J.R. 257
Berger, W.H. 212, 218
Bergin, M.J. 256
Bergstrom, R.E. 168, 233
Berkman, D.A. 88, 126
Berner, R.A. 36, 54, 56, 57, 59, 126
Berry, F.A.F. 228, 243, 244
Berry, W.B.N. 214
Bestwick, L.K. 277
Bigarella, J.J. 258, 259
Billings, G.K. 250
Binda, P.L. 130
Bingham, C. 64, 141
Biscaye, P.E. 46, 104, 126, 159
Bishop, R.S. 241
Bitterli, P. 194, 260
Björlykke, K. 51, 113, 117, 127, 189, 248
Blatt, H. 14, 127, 144, 188, 194, 248
Blissenbach, E. 64, 127, 268
Boersma, A. 107, 131
Bohor, B.F. 194
Borradaile, G.J. 58, 127
Borrello, A.V. 259

Subject Index

Numbers in italics refer to figures and tables.

Index of Stratigraphic Units

These are *not* included in subject index. Italics indicate a figure. Geologic formations of Figure 2.13 are not indexed.

Abbreviations of Rocks and Fossils*

@	At	bri	Bright
abnt	Abundant	brit	Brittle
abv	Above	brd	Bored
acic	Acicular	brn	Brown
Alg	Algae (al)	bulb	Bulbous
amor	Amorphous	bur	Burrowed
ang	Angular		
anhed	Anhedral	c	Coarse (ly)
anhy	Anhydrite (ic)	calc	Calcite (areous)
app	Appear	carb	Carbonaceous
aprox	Approximate (ly)	Casph	Calcisphaera
arg	Argillaceous	cbl	Cobble !64-256 mm)
argl	Argillite	Ceph	Cephalopod
ark	Arkose (ic)	cgl	Conglomerate
asph	Asphalt (ic)	chal	Chalcedony
		Chara	Charophytes
bar	Barite (ic)	chit	Chitin (ous)
bd (d)	Bed (ded)	chk	Chalk (y)
bdeye	Birdseye	chlor	Chlorite
bdg	Bedding	cht (y)	Chert (y)
bent	Bentonite (ic)	Chtz	Chitinozoa
bf	Buff	cl	Clastic
biocl	Bioclastic	clr	Clear
bioturb	Bioturbated	clus	Cluster
bit	Bitumen (inous)	cly	Clay (ey)
bl	Blue (ish)	clyst	Claystone
bldr	Boulder!(256 mm +)	cmt	Cement (ed)
blk	Black	cncn	Concentric
blky	Blocky	col	Color (ed)
bnd	Band (ed)	com	Common
Brac	Brachiopod	conc	Concretion (ary)
brgh	Branching	conch	Conchoidal
brec	Breccia (ted)	Cono	Conodont

*Slightly condensed from a list of the American/Canadian Stratigraphic Company (Denver, Colorado and Calgary, Alberta) and published by their kind permission.

coq	Coquina	gil	Gilsonite
Cor	Coral	gl	Glass (y)
crbnt	Carbonate	glau	Glauconite (ic)
Crin	Crinoid (al)	*Glob*	*Globigerina*
crpxl	Cryptocrystalline	glos	Gloss (y)
ctd	Coated	gn	Green
ctc	Contact	gr	Grain (ed)
cvg	Cavings	gran	Granular
deb	Debris	Grap	Graptolite
decr	Decrease (ing)	grd	Grade (ed)
dend	Dendrite (ic)	grdg	Grading
dess	Desiccation	grnl	Granule
dism	Disseminated	grnt	Granite
dk	Dark (er)	grnt.w	Granite wash
dns	Dense (er)	gsy	Greasy
dol	Dolomite (ic)	gy	Gray
dolst	Dolostone	gyp	Gypsum (iferous)
drsy	Druse (y)	gywk	Graywacke
dtrl	Detrital (us)		
		hd	Hard
Ech	Echinoid	hem	Hematite (ic)
elg	Elongate	hex	Hexagonal
Endo	*Endothyra*	hrtl	Horizontal
euhed	Euhedral	hvy	Heavy
		hydc	Hydrocarbon
f	Fine (ly)		
fau	Fauna	ig	igneous
Fe	Iron-Ferruginous	imbd	Imbedded
Fe-mag	Ferro-magnesian	imp	Impression
Fe-st	Ironstone	incl	Included (sion)
fib	Fibrous	incr	increase (ing)
fis	Fissile	ind	Indurated
fld	Feldspar (thic)	indst	Indistinct
flk	Flake (y)	intbd	Interbedded
flor	Fluorescence	intcl	Intraclast (s)
flt	Fault (ed)	intfrag	Interfragmental
ltg	Floating	intgran	Intergranular
fnt	Faint (ly)	intgwn	Intergrown
Foram	Foraminifera	intlam	Interlaminated
fos	Fossil (iferous)	intpt	Interpretation
fr	Fair	intstl	Interstitial
frac	Fracture (ed)	intv	Interval
frag	Fragment (al)	intxl	Intercrystalline
fri	Friable	ireg	Irregular
frmwk	Framework	irid	Iridescent
fros	Frosted		
Fus	Fusulinid	kao	Kaolin
g	Good	lam	Laminated
Gast	Gastropod	lchd	Leached

len	Lentil (cular)	p	Poor (ly)
lig	Lignite (ic)	pbl	Pebble
lith	Lithographic	pel	Pellet
lmn	Limonite (ic)	perm	Permeability
lmpy	Lumpy	pet	Petroleum (iferous)
lmy	Limy	phos	Phosphate (ic)
lrg	Large (er)	piso	Pisolite (ic)
ls	Limestone	pit	Pitted
lstr	Lustre	pk	Pink
lt	Light (er)	plas	Plastic
		Plcy	Pelecypod
m	Medium	pl	Plant
magn	Magnetic	pity	Platy
magnt	Magnetite	pol	Polish (ed)
mar	Maroon	por	Porous (sity)
mas	Massive	pos	Possible (ility)
mat	Material, matter	p-p	Pin point
meta	Metamorphic	pred	Predominant (ly)
mica	Mica (eous)	pres	Preserved (ation)
mic	Micro	prim	Primary
mnr	Minor	pris	Prism (atic)
mnrl	Mineral (ized)	prly	Pearly
mnut	Minute	prob	Probable (ly)
Mol	Mollusca	prom	prominent (ly)
mot	Mottled	psdo	Pseudo
mrlst	Marlstone	pt	Part (ly)
mrly	Marly	ptch	Patch (es)
msm	Metasomatic	ptg	Parting
mtx	Matrix	pyr	Pyrite (ic (ized)
musc	Muscovite	pyrbit	Pyrobitumen
n	No, none	qtz (c)	Quartz (itic)
nod	Nodule	qtzs	Quartzose
num	Numerous	qtzt	Quartzite
o	Oil	rad	Radiate (ing)
och	Ochre	rd	Round (ed)
od	Odor	repl	Replaced (ing) (ment)
olvn	Olivine	resd	Residue (al)
onc	Oncolites	rexl	Recrystallize (ation)
ooc	Oocast (ic)	rhmb	Rhomb (ic)
ool	Oolite (ic)	rmn	Remains (nant)
oom	Oomold (ic)	rr	Rare
op	Opaque	rsns	Resinous
org	Organic	rthy	Earthy
orng	Orange		
orth	Orthoclase	s	Small
Ost	Ostracod	sa	Salt
ovgth	Overgrowth	sa-c	Salt cast (ic)
ox	Oxidized	S	Sulphur

sat	Saturated		sy-Ca	Sparry calcite
sb	Sub		sz	Size
sc	Scales			
scat	Scattered		tab	Tabular
sch	Schist		*Tas*	*Tasmanites*
sd	Sand (1/16-2 mm)		*Tent*	*Tentaculites*
sdy	Sandy		tex	Texture
sec	Secondary		thk	Thick
sed	Sediment (ary)		thn	Thin
sel	Selenite		thru	Throughout
sept	Septate		tr	Trace
sft	Soft		trip	Tripoli (ic)
sh	Shale		trnsl	Translucent
shy	Shaly		trnsp	Transparent
sid	Siderite (ic)		tt	Tight (ly)
sil	Silica (eous)		tub	Tubular
sks	Slickensided		tuf	Tuffaceous
sl	Slight (ly)			
sln	Solution		uncons	Unconsolidated
slky	Silky		unident	Unidentifiable
slt	Silt			
sltst	Siltstone		v	Very
slty	Silty		var	Variable
sm	Smooth		vcol	Varicolored
sol	Solitary		ves	Vesicular
sp	Spot (ted) (ty)		vgt	Varigated
spec	Speck (led)		vit	Vitreous
Spg	Sponge		vn	Vein
sph	Spherules		volc	Volcanics
sphal	Sphalerite		vrtl	Vertical
spic	Spicule (ar)		vrvd	Varved
spl	Sample		vug	Vug (gy) (ular)
splty	Splintery			
Spr	Spore		/	With
srt	Sort (ed) (ing)		w	Well
ss	Sandstone		wh	White
stmg	Streaming		wk	Weak
stn	Stain (ed) (ing)		wthrd	Weathered
str	Streak		wtr	Water
strg	Stringer		wvy	Wavy
stri	Striated		wxy	Waxy
Strom	Stromatoporoid			
stromlt	Stromatolite		xbd	Cross-bedded (-bedding)
struc	Structure		xl	Crystal (line)
styl	Stylolite (ic)		xlam	Cross-laminated
Stylio	*Styliolina*			
suc	Sucrosic		yel	Yellow
sug	Sugary		zeo	Zeolite
surf	Surface		zn	Zone